SpringerBriefs in Environmental Science

More information about the series at http://www.springer.com/series/8868

Linda Gonçalves Veiga · Mathew Kurian
Reza Ardakanian

Intergovernmental Fiscal Relations

Questions of Accountability and Autonomy

UNITED NATIONS
UNIVERSITY

UNU-FLORES
Institute for Integrated Management
of Material Fluxes and of Resources

Springer

Linda Gonçalves Veiga
NIPE
University of Minho
Braga
Portugal

Mathew Kurian
Reza Ardakanian
UNU-FLORES United Nations University
Dresden
Germany

ISSN 2191-5547 ISSN 2191-5555 (electronic)
ISBN 978-3-319-06295-2 ISBN 978-3-319-06296-9 (eBook)
DOI 10.1007/978-3-319-06296-9

Library of Congress Control Number: 2014946405

Springer Cham Heidelberg New York Dordrecht London

Printed on acid-free paper

Springer is part of Springer Science+Business Media (www.springer.com)

Acknowledgements

Linda Veiga would like to thank Francisco Veiga for his collaboration in writing module III and Henry Chappell for helpful comments and suggestions on modules I and II.

Contents

Figures

Tables

Chapter 1
Trends in Financing of Public Services

Abstract The first section of the volume is dedicated to trends in financing of public services. It starts by discussing the rationale for government intervention in the economy, and by presenting an overview of the changing role of government over time and across countries. It then presents a picture of government architectures across countries, and discusses normative and positive issues that influence the allocation of functions to different levels of government and the sharing of revenues across them. Finally, in the third section, the major trends in decentralisation are discussed and case studies of the impact of decentralisation on public service delivery are reviewed.

Keywords Decentralisation · Government architectures · Government functions · Impact of decentralization · Intergovernmental fiscal transfers

The first section of the volume provides the necessary framework to understand fiscal interrelations between different levels of government. It starts by discussing the normative issues related to government intervention in the economy, and by presenting an overview of the changing role of government over time and across countries. The second section presents a global picture of decentralisation around the world and over time. It is also dedicated to the normative and positive issues that influence the allocation of functions to different levels of government and the design of intergovernmental fiscal transfer mechanisms. Finally, in the third section, the major trends in decentralisation are discussed. Over the last decades, important and influential decentralisation reforms have taken place worldwide. Especially in the developing world, this trend towards decentralisation has been supported by international agencies and by the results of extensive literature on this topic. The last part of the section provides an overview of the attempts that have been made to evaluate empirically the impact of decentralisation on the quality of public goods delivery, focusing on water and sanitation; education and health; and poverty, income inequality and regional disparities.

L.G. Veiga et al., *Intergovernmental Fiscal Relations*,
SpringerBriefs in Environmental Science, DOI 10.1007/978-3-319-06296-9_1

1 Government Intervention in the Economy

In any country, the role of government in the economy is subject to much debate. Some believe that the government does too little; while others think that, it has grown too much. Although the balance between the public and the private sectors will always be a controversial issue, there is little dispute that for markets to work properly there must be a government to define property rights and enforce contracts. Without rules defining the ownership of property, it would not be possible to prohibit theft. Therefore, enterprise would be discouraged, as potential gains from trade would be appropriated by others. Contracts regulate exchange; they ensure that contracting parties carry out the tasks they have agreed to do. To guarantee property rights and enforce contracts, the government needs to operate a police service and a court system. However, for this, only a small government with low taxes is needed. So the question becomes: should the government provide minimal functions and otherwise leave markets alone?

1.1 Why Should the Government Intervene in the Economy?

The ultimate goal of government intervention in the economy is to promote the economic well-being of citizens by improving efficiency and equity (fairness). The efficiency goal refers to public policies designed to promote the best use of economic resources. The equity side of public policy design is associated with redistributive policies intended to reduce inequalities in the distribution of economic benefits.

Markets often fail, leading to inefficiencies in the allocation of resources. For example, when an individual's economic activity generates externalities (effects imposed by one economic agent on another without its consent), inefficiencies will arise that warrant government intervention in the economy. Environmental resources, such as water, soil and air, are frequently used for disposal of wastes without taking into account the negative externalities generated or the benefits they might have in alternative uses. This happens because no one owns environmental resources. Therefore, in the absence of government intervention, polluters do not have to pay for the right to use them.

Other examples of market failures include imperfect competition and the existence of public goods. Imperfect competition can take many forms, but a typical consequence is that producers limit output to create a scarcity, which results in higher price and lessened gains from trade. Public goods have two characteristics that, in combination, imply that they will not be efficiently provided in private markets. First, public goods are non-rival in consumption, implying that one person's consumption does not diminish the amount of the good available for consumption by others. Second, public goods are non-excludable, implying it is difficult to exclude somebody from consuming a good if they do not pay for it.

Examples of public goods are national defence, lighthouses, flood control systems and street lighting. Due to their characteristics, these goods will either not be supplied by the market or they will be supplied in an insufficient quantity.

Besides market failures, government intervention in the economy is also motivated by inequality in the distribution of income, wealth or opportunities resulting from free market outcomes. Public policies to promote equity are often found in social security, education and health sectors. Access to safe and clean drinking water and sanitation influences children's development and the health conditions of people, so the provision of water and sewage to the population can also be justified by equity goals.

Efficiency and equity goals are interconnected. Policies promoting efficiency can increase inequality, while redistributive policies are likely to influence the choices of both recipients and taxpayers (namely the incentives to work and spending decisions) in ways that may cause losses of efficiency. However, and especially in developing countries, the trade-offs between efficiency and equity may be less pronounced. Inequity and poverty exacerbate degradation of environmental resources (Dasgupta et al. 2005; Lal 2008). Thus, public policies to reduce poverty and promote economic development may enhance efficiency and alleviate environmental problems. On the other hand, policies promoting sustainable use of natural resources, which take into account the water, soil and waste nexus, will enhance human well-being not only for the present generation but also for future generations. In fact, global food security is highly dependent on finite resources, such as soil and water, which are prone to degradation and pollution by misuse and mismanagement.

As discussed above, markets often fail to produce fully efficient outcomes. However, government intervention does not always improve what the unregulated economy can achieve. Reasons for government failure include lack of information, difficulties in predicting and controlling market responses, problems managing bureaucracies, corruption and political constraints.[1] The appropriate role of government in the economy has to take into account both market and government imperfections. Over time and across countries, different views have prevailed about the desirable boundaries of government intervention.

1.2 The Size and Growth of Government

To measure government intervention in the economy several indicators can be used; specifically, the weight of public expenditure in Gross Domestic Product (GDP), the weight of public revenues in GDP or the percentage of public employees in the work force. The most frequently used indicator is the percentage of public expenditure in GDP. It is a broad measure of the scope of government activity in the

[1] Some of these issues will be discussed in the third section of this volume.

economy and allows for comparisons over time and among countries since it is measured relative to GDP.

Although different countries have followed different paths, during the twentieth century there has been remarkable growth in the size of government in many economies (Tanzi and Schuknecht 2000; Tanzi 2005). Figure 1 shows the growth of general government expenditures as percentage of GDP, from 1870 to 2002, in six countries (France, Germany, Japan, Sweden, the United Kingdom and the United States of America).

In the nineteenth century, government provided minimal economic functions. The increase in military expenses associated with the two world wars, and the great depression during 1929–1933 are the major events explaining the growth in the size of government over the first half of the twentieth century. But the most remarkable increase in government intervention in the economy occurred between 1960 and 1980. During this period, most countries were not involved in wars, there was no severe depression and demography was fiscally friendly. Growth was due to an unprecedented enthusiasm for the positive role of the state in the allocation of resources, stabilisation of the economy and promotion of equity. However, governments had little success with economic stabilisation during the stagflation of the 1970s. This failure, along with rapid growth of spending, taxation and public debt, tempered optimism about government's role in the economy. Beginning in the early 1980s, several countries tried to reduce the size of government, but realised that it was easier to increase it than to reduce it.

Figure 2 shows the path of government spending from 1991 to 2013 for the same six countries, plus an average for emerging market and developing economies. In five of the developed countries considered, the size of government stabilised or even increased. Only Sweden managed to reduce it significantly from 71 % in 1993 to 53 % in 2013. When comparing developed economies with emerging market and developing economies, it is clear that government size is considerably lower in the second group of countries, although it grew from 25 % of GDP in 2001 to 30 % in 2013. The effect of the current global financial and economic crisis is also visible in

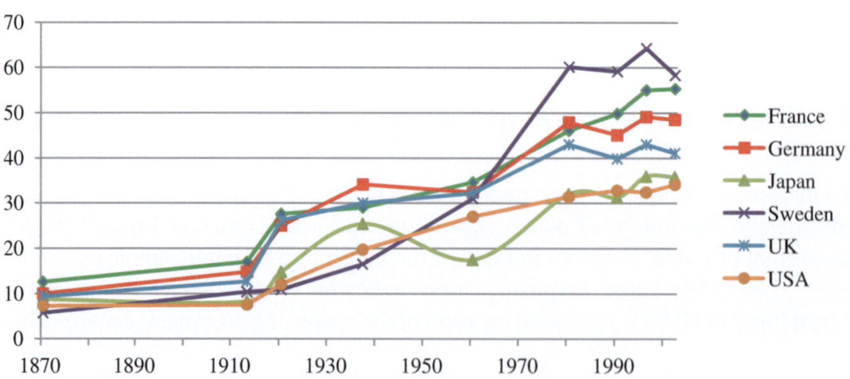

Fig. 1 General government expenditure, 1870–2002 (% GDP). *Source* Tanzi (2005)

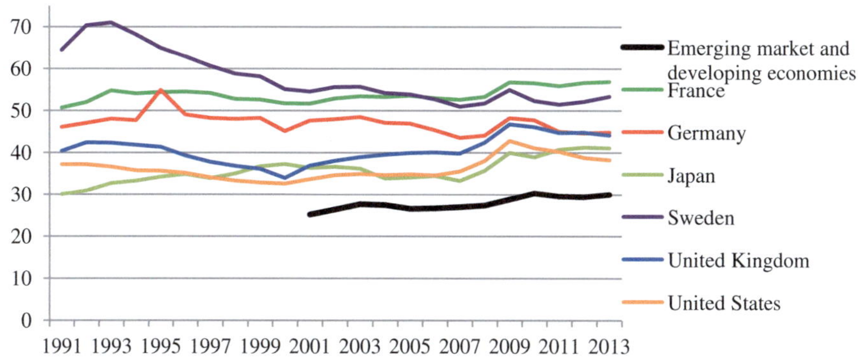

Fig. 2 General government expenditure, 1991–2013 (% GDP). *Source* IMF (2013b). *Note* Emerging market and developing economies is based on an average of 154 countries

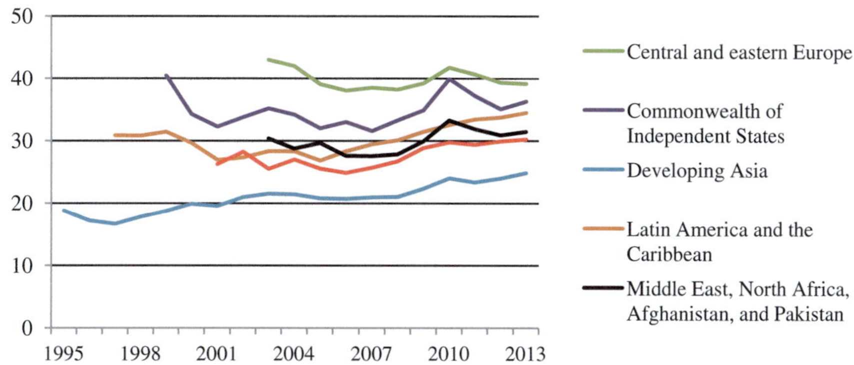

Fig. 3 General government expenditure, 1991–2013, emerging markets and developing economies (% GDP). *Source* IMF (2013b), Central and eastern Europe refer to an average of 14 countries; developing Asia, 29 countries; Latin America and the Caribbean, 32 countries; Middle East, North Africa, Afghanistan and Pakistan, 22 countries; and Sub-Saharan Africa, 45 countries. For further details please refer to IMF World Economic Outlook Databases web page

the picture, with an increase in expenditure levels that reached its peak in 2009, followed by a reduction in most of the countries.

When considering subgroups of countries from emerging markets and developing economies (Fig. 3), a diversity of perspectives about the role of the state in the economy becomes clear. Asian developing countries tend to have the lowest levels of public expenditure, while central and eastern European countries have the highest. As in the previous graph, the effects of the current global crisis are also visible, but the peak occurred one year later, in 2010.

2 Multi-level System of Government

An important issue in the public finance literature is the appropriate division of responsibilities for managing public functions and resources across different levels of government.[2] The topic is particularly relevant for developing and emerging countries, where well-designed decentralisation reforms have a higher potential to improve efficiency in the provision of public services, and to enhance the development of integrated and sustainable strategies for the use of water, soil and waste resources.

In this subsection, we present a picture of government architectures across countries, and discuss the normative issues related to the assignment of functions to different levels of government, and the sharing of revenues across different levels of governments. Finally, we present a short digression on political economy issues related to decentralisation that are particularly relevant for the design of intergovernmental fiscal transfer mechanisms.[3]

2.1 Government Architectures

There is great diversity in government architectures across the world. The number of subnational levels of government varies considerably from country to country, as well as the number of jurisdictions in each tier of government.[4] According to Gómez-Reino and Martinez-Vazquez (2013), in a sample of 197 countries, more than half (101) have two levels of subnational governments, 50 countries have three levels, 35 have one and only 10 have four tiers of subnational government. The number of jurisdictions per level of government also varies widely in the sample. The two extreme cases are Kiribati with no local government and India with more than 240,000. The median value is 194 jurisdictions per level of government.

Countries' government architectures evolve over time and reorganisational reforms are usually subject to intense political debate. When analysing the main drivers of jurisdictional fragmentation, Gómez-Reino and Martinez-Vazquez (2013) conclude that the number of tiers of government is strongly positively related to both population and land area size. Regarding the size (measured by area and population) and number of local governments, their results indicate that: a higher

---

[2] For surveys on fiscal federalism see Oates (1999) and Ahmad and Brosio (2006). For analysis of fiscal decentralisation management refer to Ahmad and Tanzi (2002), Smoke (2006), Boadway and Shah (2007), Shah (2008), and Fedelino and Ter-Minassian (2010).

[3] The third section of the volume is entirely dedicated to the issues of accountability and policy outcomes.

[4] For country profiles see the Global Observatory on Local Democracy and Decentralization from the United Cities and Local Governments' at http://www.cities-localgovernments.org/gold/country_profile.asp.

number of tiers of government lead to jurisdictional fragmentation; population and land area are positively correlated with jurisdictional fragmentation; preferences for political accountability increase the number of governments; and finally variables measuring heterogeneity of preferences are also influential.

As far as the functions that are assigned to different levels of government, typically, social protection (e.g. pensions and unemployment benefits) and general public goods and services (e.g. law and order, national defence, national infrastructures) are provided by the central government. Among others, subnational governments have responsibilities in education, health, environmental protection, housing and community amenities (e.g. distribution of water and electricity, sewage) and recreation, culture and religion. The question then becomes, what are the economic principles that ought to guide the structure of governments' architecture and the assignment of functions to different levels of governments?

2.2 *The Assignment of Functions Among Levels of Government*

The main economic argument for decentralisation is that subnational governments have more proximity with the population, and therefore, they can better match the provision of public goods and services to the needs of the population. The more heterogeneous the preferences of different communities, the larger the gains that can be obtained through decentralisation.[5] Other issues that need to be taken into account are the existence of economies of scale[6] in production and internalisation of externalities. In the presence of economies of scale, excessive fragmentation in the provision of public goods results in higher costs and it may be beneficial to organise the provision at a higher government level. Additionally, when public policies adopted by a subnational government generate externalities to neighbouring political jurisdictions (e.g. police protection or reduction in pollution), centralisation is beneficial because central governments are likely to internalise spillovers, while subnational governments are not. In this case, unless incentive schemes are designed by the central government, local decision makers may only consider the costs and benefits to their constituencies resulting from the public good.

Political concerns also justify decentralisation since it may create spaces for democratic representation and increase the accountability of governments to citizens. Decentralised decision-making allows each jurisdiction to communicate its needs and preferences to the central government. On the other hand, proximity between local politicians and voters allows voters to discern better whether the politicians' actions are in accordance with their interest, and to punish or reward

---

[5] Using a large sample of countries, Shelton (2007) concludes that heterogeneity in preferences (measured by an index of ethno-linguistic fractionalisation) increases decentralisation.

[6] Economies of scale occur when average costs decrease as the output produced increases.

them accordingly at elections. In Latin America, decentralisation played an important role in the democratisation process some countries have undergone. In Africa, the end of long civil wars and the spread of multiparty political systems led to the transfer of decision-making to lower levels of government. In former socialist states, the need for decentralisation resulted from the collapse of central planning.

However, decentralisation, and particularly the creation of additional subnational governments may bring additional costs. Costs are associated with new administrations (buildings, staff, equipment), with local elections, and extra coordination needs. In certain cases, the joint delivery of services, for example of solid waste collection and water supply, by a higher level of government or by consortiums of subnational governments, may generate savings.[7]

2.3 *Efficiency and Equity Issues with Mobile Resources*

In order to serve their constituency, subnational governments choose to offer different sets of public goods and charge different local taxes. If individuals have geographic mobility, they will choose to live in the jurisdiction that better serves their preferences in terms of local public goods and services, and local taxes. Mobility may increase the gains from decentralisation by creating homogeneous communities (citizens choose to live in jurisdictions where residents have preferences similar to theirs) and because local governments' competition to attract population may enhance efficiency.

Mobility is also a relevant issue when it comes to the promotion of equity. If redistributive functions are mainly assigned to local governments, poor households will move to jurisdictions where social transfers are high and taxes low, while rich households will do the opposite. In the absence of coordination, policies will become unsustainable. Therefore, mobility of households requires coordination in redistributive policies. Another argument for the assignment of this function to the central government is that the equity pattern should be national and not local, in order to avoid inequality among individuals living in different geographic areas of a country.

Capital is highly mobile, even more than people are. It moves easily across geographic areas within a country, and even across national borders, opening up additional needs for coordination. If capital is perfectly mobile, an increase in local taxes on capital diminishes the supply of capital since it moves to other jurisdictions offering better conditions. Therefore, subnational governments may engage in an inefficient race to the bottom, setting lower tax rates to retain capital. They also have an incentive to reduce taxes on capital and raise the taxes on less mobile workers, which creates other problems.

[7] Callan and Thomas (2001) analysed municipal waste services and concluded that the joint provision of recycling and disposal services generates cost savings.

2.4 Intergovernmental Fiscal Transfers

In general, the central government is responsible for collecting the majority of public funds, and for transferring resources to lower levels of government.[8] Lower levels of government have more limited taxation powers. In transition economies and developing countries, intergovernmental transfers represent about 60 % of subnational government expenditures, while in member countries of OECD, they represent about one-third.

2.4.1 Purposes of Transfers

Revenues may be most efficiently collected by the central government, but at least in part, most efficiently spent at the subnational level. Overall efficiency then requires revenue sharing between governmental levels in order to achieve the correct balance between responsibilities and resources at each governmental level (vertical fiscal balance).

In addition, revenue sharing can be used to transfer resources to communities who need them but lack the fiscal capacity to generate adequate revenue. To provide the necessary amount of local public goods, these jurisdictions would otherwise have to impose higher tax burdens to the population. Intergovernmental transfers can be used as fiscal equalisation schemes to promote equity among jurisdictions (horizontal financial balance).

Finally, the existence of spillovers in the provision of public goods requires centralisation or intergovernmental transfers to induce the right level of provision. Spillovers occur when policies adopted by a subnational government affect not only the well-being of people in their jurisdiction, but also of residents of other communities. In short, positive or negative externalities may arise. If each subnational government cares only about its members, externalities are not taken into account, and resources are allocated inefficiently. The central government should transfer money to the subnational governments in a way that provides incentives for them to undertake activities that produce positive externalities.

2.4.2 Types of Transfers

Intergovernmental transfers (or grants) can be classified as unconditional, when they may be used according to the wishes of the recipient government and, conditional when certain restrictions must be met by the recipient governments. Restrictions can be input-based or output-based. In the former case, the type of

[8] For an overview of intergovernmental fiscal transfers, see Geys and Konrad (2010), for a practitioner's guide, see Shah (2007). Bird and Smart (2002) provide international lessons for developing countries.

spending that grants can be used for is imposed; while in the latter, transfers require the achievement of certain results in service delivery. In principle, output-based transfers by linking results with transfers, induce higher responsibility in local management and accountability for results to citizens.

Horizontal disparities in income or wealth across jurisdictions can justify unconditional equalisation transfers (based on fiscal capacity and expenditure needs) that allow the recipient government to decide the best way to spend them. Unconditional transfers also play an important role in correcting vertical imbalances, namely when subnational governments lack funds to perform the responsibilities assigned to them. They increase the recipients' resources (income effect) and preserve local autonomy, thereby enhancing local welfare. Empirical studies show that the main disadvantages of general-purpose transfers are that they reduce local authorities' incentives to promote economic growth (Weingast 2014) and increase local public spending (Bucovetsky and Smart 2006). This last phenomenon is known as 'flypaper effect'. For political and bureaucratic reasons, money received by the community in the form of unconditional transfers results in greater public spending than if the money had been transferred directly to local residents. The tax relief resulting from transfers that could be used by local residents to spend on private goods is smaller than expected.

Unconditional grants used to provide broad support in an area of subnational expenditures (e.g. education, environmental protection) are called block transfers. In this case, recipients only have discretionary power to decide the specific uses of the funds within the broad area of performance.

When the central government intends to stimulate spending on a certain public good (influence local preferences), it is more efficient to use conditional (or specific purpose) transfers. This may be the case when a public programme generates strong positive externalities to neighbouring jurisdictions that the local government does not take into account. By subsidising the programme, the donor government internalises the externalities. With non-matching conditional transfers, the donor stipulates where the money is to be spent but does not require local matching. Conditional transfers often take the form of matching transfers: a certain proportion of a specific project is financed by the donor, but must be supplemented with outlays of the recipient governments. By reducing the cost of the project to the local community, the transfer stimulates its supply. Matching transfers induce more scrutiny and responsibility to the subnational government. To put a ceiling on the cost, the granting government may even specify the maximum amount it will contribute. In that case, we have a matching closed-ended transfer.

2.4.3 Transfer System Designs

The settlement of transfers, particularly fiscal equalisation grants, is usually supported by formulas, which use indicators of both the needs of the population and of the local fiscal capacity.

To measure fiscal capacity, that is the ability of a subnational government to raise revenues from own sources, two approaches can be used. The first is to use macro measures of the jurisdiction (e.g. gross domestic product or income). The second approach uses information from the tax system and intends to equalise the tax base across jurisdictions using as reference the national mean or median. Transfers (*T*) to a jurisdiction *i* are determined according to the following formula:

$$T_i = POP_i^* \left(PcTb_{nat}^* t_{nat} - PcTb_i^* t_i\right),$$

where the index *nat* stands for the national mean or median, *POP* for population, *PcTb* for per capita tax base and *t* for the tax rate. T_i is negative for richer than average (or mean) jurisdictions, and positive for the poorer. The subnational level data necessary to determine transfers are usually better (in terms of accuracy and time) with the tax system approach than with the one that uses macro indicators. Despite these disadvantages, macro indicators are used in Brazil and India.

To measure the needs for local public goods and services, population is the major determinant. In addition to population, other demographic data is relevant. For example, differences in age composition of the population result in different needs for schools or support for the elderly. The same is true regarding the incidence of poverty or diseases. Population density, the geographic area of the jurisdiction, as well as its altimetric amplitude may also induce need differentials. Formulae to determine the component of intergovernmental transfers (i.e. variables taken into account and their weights) associated to expenditure needs vary considerably across countries.

2.4.4 Political Economy Issues

The first generation of studies on decentralisation assumed that governments acted like a benevolent dictator whose only objective is the maximisation of social welfare. With disperse preferences or needs among local communities, decentralisation provides subnational officials with discretionary power and autonomy that generates increase in welfare. The second generation of papers recognised that political influences could play an important role in decentralisation and emphasised the need for political accountability at all levels of government (Oates 2005; Kurian 2010; Weingast 2014).

At the central level, grants can be used as an instrument to woo the electorate before national or local elections, as suggested by the literature on pork barrel policies and political business cycles (Rogoff 1990). The donor government may also use them for tactical redistribution among jurisdictions by favouring its supporters in the allocation of funds (Cox and McCubbins 1986; Lindbeck and Weibull 1987) or targeting localities with more swing voters (Dixit and Londregan 1996). The empirical literature analysing the political determinants of the distribution of intergovernmental grants is extensive. For block grants in Albania, Case (2001) found evidence that more assistance was allocated to swing communes and to those that might be pivotal to winning a majority of seats in the Parliament. For the Mexican case, Diaz-Cayeros et al. (2007) found evidence that under the PRI (Institutional Revolutionary Party) rule municipalities supporting the opposition received lower transfers. For the Indian federation, Khemani (2007) concluded that the delegation of transfers to an independent agency reduced partisan influences in the allocation of transfers to states. Allers and Ishemoi (2011) found that the introduction of formulae for the assignment grants to regions in Tanzania did not reduce political manipulation in grant allocation. Because actual grant allocation continued not to be ruled by formulae, regions with a higher representation at the national parliament continued to receive more grants.

Political issues may also interfere with the degree of decentralisation within a country. By dominating parties, national elites may introduce institutional changes that reduce subnational governments' powers, increasing centralisation. This was the case in Mexico under the PRI's rule (Diaz-Cayeros 2007) and in India under the Congress Party (Rao and Singh 2005). On the opposite side, intergovernmental transfers may generate incentives for subnational officials to free ride on common resources, for example, by accumulating debt and augmenting the need for a bailout.[9]

Recently, several papers have noted that political factors can lead to inefficiency in the assignment of functions to subnational governments. For example, if preferences for public goods are skewed, so that the preference of the median voter is above the average preference, decisions based on a majority voting rule will lead to overproduction of the public good (Besley and Coate 2003; Lockwood 2002, 2008).

In developing countries and new democracies, where institutions are weak and rules difficult to enforce, political pressures play a stronger role in creating inefficiencies. Decentralisation may exacerbate the conflict in divided societies, as highlighted by Eaton (2006) for the Columbian case. However, there are examples in which decentralisation fosters democracy. In Taiwan, the transition from an authoritarian regime to a democracy was initiated by small reforms that began at the local level (Weingast 2014). Myerson (2006) points out that decentralisation provides opportunities for local politicians to acquire experience and reputation that may be useful in national elections. In countries with strong ethnic, cultural or linguistic differences (e.g. Belgium, India, Spain, the Netherlands, Indonesia, the Philippines and South Africa), decentralisation reduces conflicts and helps to maintain the unity of the country (Stepan 2004; Inman and Rubinfeld 2012).

[9] This topic will be addressed more extensively in the second section of the volume.

3 Worldwide Trends in Decentralisation

Over the last decades, important influential decentralisation reforms have taken place worldwide. At least in the developing world, this trend towards decentralisation has been supported by international agencies, such as the World Bank and the United Nations, and by the results of extensive literature on the topic.

3.1 Global Picture

3.1.1 Decentralisation Around the World

Decentralisation, the transfer of power and resources from the central government to subnational governments, is a complex concept involving fiscal, political and administrative dimensions. Therefore, measuring decentralisation is also complex.[10] International organisations have made important efforts to collect data on several dimensions of decentralisation. On their web pages, the OECD makes available Fiscal Decentralization Databases, the World Bank produces Fiscal and Political Decentralization Indicators, and the United Cities and Local Governments have a Global Observatory for Local Democracy and Decentralization. Recently, Ivanyna and Shah (2014) developed a unique aggregate index of decentralisation dataset for 182 countries, which takes into account the relative importance of local government, the security of existence of local governments, and fiscal, political and administrative indices of decentralisation.[11] Figure 4 shows the world picture of the aggregate index. Developed countries are generally more decentralised, while African countries are the most centralised, Latin American and Asian countries are in between.

For comparisons of decentralisation over time, we rely on the World Bank's fiscal and political decentralisation indicators available on the web. Political indicators are quite limited regarding the variables used, so we focus on the fiscal indicators. These indicators use data from the Government Finance Statistics (GFS) of the International Monetary Fund, which is a good source for panel data analysis of fiscal flows, even though it has limited coverage across developing economies. Ten indicators of decentralisation are presented, grouped in four categories: subnational revenues, subnational expenditures, subnational functional expenditure and fiscal gap. We concentrate on the share of subnational governments' expenditures in consolidated general government expenditures for all levels of government.

[10] See Abdelhak et al. (2012) for an overview of attempts to measure and collect data on decentralisation around the world.

[11] The paper provides a detailed explanation of the aggregate index, and tables with values of the index per country.

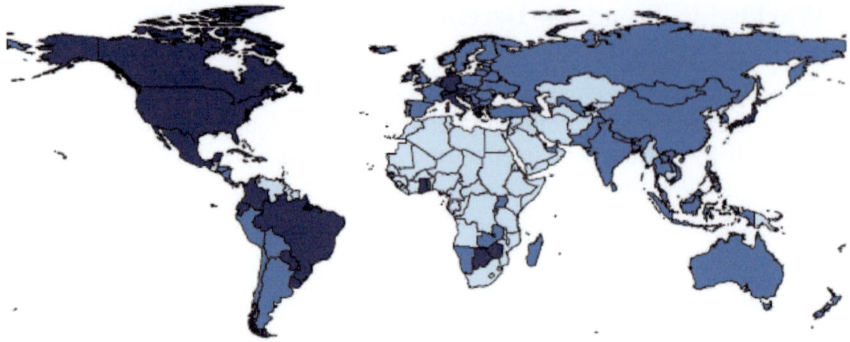

Fig. 4 World map—Ivanyna and Shah (2014) decentralisation index. *Note*: Shades of the colour correspond to 0–12th, 25–50th, 50–75th, 75–100th percentiles of index of decentralisation. *Source* Ivanyna and Shah (2014, p. 21)

Table 1 provides 5-year averages of this indicator, starting in 1975 until 2010, for several countries.

Latin American countries decentralised mostly during the 1980s and 1990s.[12] Decentralisation was part of the democratisation process, as autocratic central regimes were replaced by elected governments. In Africa, the majority of countries have adopted some decentralisation reforms over time, particularly in recent years. However, countries have progressed at different rates and Africa remains the least decentralised continent (recall Fig. 1). Political changes resulting from the end of long civil wars, an increase in the number of multiparty political systems, and pressures from regional or ethnic groups for more autonomy, have increased the demand for decentralisation. For Africa, Table 1 reports data for South Africa (the only country available), where a significant increase in subnational governments' share of public expenditures occurred during the period considered. In most countries of the Asia-Pacific region, the degree of decentralisation is more stable. However, some countries (particularly in East Asia) have adopted decentralisation measures in order to improve service delivery to large populations. Transition countries implemented decentralisation plans as the former socialist way of organising economic activities collapsed. Even in industrialised countries, significant progress towards decentralisation was achieved in some countries, namely Italy, Spain and the USA.

3.1.2 Decentralisation by Functions

As can be seen from Table 2, the degree of decentralisation by function differs quite substantially among countries. Subnational governments play an important role in a large range of services in many countries, particularly in education, health, housing

[12] For a report on decentralisation and local democracy all over the world, see United Cities and Local Government (2010).

Table 1 Share of subnational government expenditure in total public expenditure

	Cash/Accrual[a]	1976–1980	1981–1985	1986–1990	1991–1995	1996–2000	2001–2005	2006–2010
Latin America								
Bolivia	Cash/Accrual			12	13	23	20/21	26
Brazil	Cash			27	30	39		36
Mexico	Cash			14	20	22		
Peru	Cash			7	17	17	20	21
Africa								
South Africa	Cash		25	22	16	23	50	
Asia								
Indonesia	Cash	11	11	13	12	10		
India	Cash	37	38	37	38	39	39	41
Thailand	Cash				5	6	6	6
Transition countries								
The Czech Republic	Cash				15	17	18	19
Hungary	Cash		17	17	21	21		
Romania	Cash/Accrual	16	19	9	10	11	17/21	21
Slovenia	Cash				10	10	16	15
The Slovak Republic	Cash					8	13	18
Industrialised								
Australia	Cash/Accrual	34	35	37	38	38/39	37	38
Canada	Cash			48	49	51	52	53
Denmark	Cash/Accrual	46	43	44	44/44	45/46	49	48

(continued)

Table 1 (continued)

	Cash/Accrual[a]	1976–1980	1981–1985	1986–1990	1991–1995	1996–2000	2001–2005	2006–2010
France	Cash/Accrual	13	14	14	15/16	17	17	19
Germany	Cash/Accrual	34	34	34	33/30	35/34	35	35
Italy	Accrual				21	23	27	27
Japan	Accrual						47	47
Portugal	Accrual				10	11	12	14
Spain	Cash/Accrual	9	15	21	27/27	28/30	39	41
The United Kingdom	Cash/Accrual	24	23	24	21/22	21/23	23	23
USA	Accrual	25	24	27	36	39	40/39	35

[a] The difference between cash basis accounting and accrual accounting results from the time gap that frequently exists between the moment debits or credits appear and the moment they are paid or received by the public administration. Governments using cash basis assign revenues and expenditures to the period during which the cash is expected to be received or disbursed. Because cash basis allows governments to balance budgets by artificially delaying payments and anticipating revenues, most organisations recommend the use of accrual basis. This accounting method reports income when earned and expenses when incurred, irrespective of when they are actually received or paid

Source Own calculations based on the fiscal decentralisation indicators of the World Bank (2013)

Note For the Slovak Republic, the source was OECD (2013)'s Fiscal Decentralization Database

and environmental protection. In several countries, they are responsible for a large majority of expenditures on these functions. The least decentralised function is social protection. These go in line with the normative view that income redistribution is better achieved at the central level while services that require tailoring expenditures to local preferences should be provided by subnational governments.

The global financial and economic crisis, rapid urbanisation and demographic changes, global environmental challenges /climate change, and economic globalisation and regional integration represent important challenges for subnational governments that increase demands on them and are likely to make their role more difficult. Given the major trend of decentralisation it is, therefore, important to learn from previous experience and assess the impact of decentralisation on the quality of service delivery.[13]

3.2 Case Studies of the Impact of Decentralisation on the Quality of Service Delivery

Subnational governments are judged on their ability to provide citizens with quality public services, such as water, sanitation, health, education, waste management, transport and energy. Improving the provision of these services is key to the achievement of the United Nations Millennium Development Goals. In what follows we provide a brief overview of the attempts that have been made to evaluate empirically the impact of decentralisation on the quality of service delivery.[14]

3.2.1 Water and Sanitation

There is a trend towards decentralisation in the water sector worldwide. In Latin America, the scale of water service reform has been truly impressive and unparalleled in comparison to any other part of the world (Foster 2005, p. 29). Political pressure for decentralisation across all areas of government and growing dissatisfaction with the performance of national monopolies in water industries generated a movement towards decentralised control that began in the 1980s. In most countries, the reform has not resulted in a move from public to private provision, but rather from unregulated centralised public provision to a regulated decentralised one. Several research papers have tried to disentangle the effects of these reforms. For the city of Porto Alegre in Brazil, Santos (1998) found that, between 1989 and 1996, decentralised participatory budgeting by citizens led to increased access to basic sanitation (water and sewage) as well as enrolment in elementary or secondary schools. Faguet's (2008) results for Bolivia show that decentralisation increased the

[13] Political issues and accountability will be discussed in Sect. 3.

[14] For an extensive survey, see LDI (2013). See also Ahmad et al. (2005).

Table 2 Subnational expenditures as a percentage of total government expenditure by function, averages from 2006 to 2010

	Cash/ Accrual	General public services (%)	Defence (%)	Public order and safety (%)	Economic affairs (%)	Environmental protection (%)	Housing and community amenities (%)	Health (%)	Recreation, culture and religion (%)	Education (%)	Social protection (%)
Latin America											
Bolivia	Accrual	20	0	7	48	80	93	40	86	46	10
Africa											
South Africa	Accrual	50		85	91	87	89	96	94	94	82
Asia											
Korea	OECD Data	42	1	19	27	78	75	16	65	75	69
India	Cash	23	1	38	41		44	54		69	
China	Cash	25	2	83	70	97	93	98	89	94	94
Transition countries											
The Czech Republic	Cash	29	1	9	31	58	49	4	68	48	7
Hungary	Accrual	20	0	8	15	57	93	27	38	49	8
Romania	Accrual	25	0	4	33	72	63	1	60	59	10
Slovenia	Cash	17	1	7	25	57	67	14	47	43	2
The Slovak Republic	Accrual	16	0	3	19	55	84	0	46	64	4
Industrialised											
Australia	Accrual	24	0	84	63	80	78	49	73	58	11
Canada	Cash	41	0	68	72	79	78	82	73	95	33
Denmark	Accrual	11	1	9	44	52	43	98	54	51	57
France	OECD Data	23	0	21	39	85	85	35	74	29	30
Germany		56	0	91	61	77	76	72	93	96	41

(continued)

Table 2 (continued)

	Cash/ Accrual	General public services (%)	Defence (%)	Public order and safety (%)	Economic affairs (%)	Environmental protection (%)	Housing and community amenities (%)	Health (%)	Recreation, culture and religion (%)	Education (%)	Social protection (%)
	OECD Data										
Italy	Accrual	21	0	12	53	82	67	65	52	25	4
The Netherlands	Accrual	14	0	38	42	86	68	3	81	50	15
Portugal	Accrual	24	0	3	32	77	98	5	61	11	3
Russian Federation	Accrual	50	0	22	54	61	91	58	71	78	20
Spain	Accrual	29	0	44	58	92	96	93	81	96	11
The United Kingdom	Accrual	11	0	40	29	55	48	0	48	45	21

Source IMF (2013a)

responsiveness of public investment to local needs. After decentralisation reforms, investments in water and sanitation, water management, education and agriculture were higher where water and sewerage connection rates were lower, illiteracy rates were higher, and malnutrition was a greater risk, respectively. Based on urban/rural case studies in distinct geographical regions of Mexico, Wilder and Lankao (2006) concluded that decentralisation did not result in more efficient water management or sustainable use of water resources.

The case of rural water supply in Central India was researched by Asthana (2010), who found that the decentralised provision of water supply was less efficient. Beddies and De Soto (2006) analysed the decentralisation and privatisation of the water sector in Albania. From a demand perspective, Vásquez and Franceschi (2013) investigated household preferences for decentralisation and system reliability in water service in Nicaragua. They found that households are willing to pay more for reliable water supply and that they have more confidence in the current, centralised provider than in a decentralised service implemented at the municipal level.

Finally, using data for 83 river basins worldwide, Dinar et al. (2007) found that persistent water scarcity is a major stimulus to reform, and that the existence of dispute resolution mechanisms, greater financial responsibility of users, and government financial support to basin budget improves decentralisation performance.

3.2.2 Education and Health

The number of country analyses of the effects of decentralisation on education is quite high, and most research finds positive effects of decentralisation on enrolment ratios, responsiveness to local needs and education outcomes (such as literacy rates, years of schooling and dropout rates for primary and secondary education).[15] More recently, cross-country analyses using panels with a large number of countries (Enikolopov and Zhuravskaya 2007; Hicken and Simmons 2008) reinforced the positive effects of decentralisation on education.

For health services, evidence is mixed. Several papers found positive effects of decentralisation (namely in terms of mortality rates), while others conclude that decentralisation may generate negative effects (e.g. increases in regional disparities or reductions in the proportions of expenditures local governments assign to health).[16]

---

[15] Among others, see Faguet and Sánchez (2008) for Bolivia and Columbia, Hecock (2006) for Mexico, Simatupang (2009) for Indonesia.

[16] See, among others, Robalino et al. (2001), Khaleghian (2004), Akin et al. (2005) for Uganda and Habibi et al. (2003) for Argentina.

3.2.3 Poverty, Income Inequality and Regional Disparities

As explained in Sect. 2.3, the two main arguments for central governments to play a crucial role in redistributive policy are that equity patterns should be national, and not local, and that mobility of households requires coordination in redistributive policies, otherwise they become unsustainable. However, subnational governments play an important role in redistributive policies (Bahl et al. 2002). First, they are important players in the delivery of basic services such as health, education and welfare. Second, they have the ability to charge local taxes and user fees that have redistributional consequences. Because subnational governments are closer to the population, they have better information to target those in need. However, particularly in developing countries, decentralised antipoverty and development programmes may be captured by local elites.[17]

Therefore, the impact of decentralisation reforms on poverty, income inequality and regional disparities is not clear. Using a large sample of countries over three decades, Sepulveda and Martinez-Vazquez (2011) found evidence that fiscal decentralisation tends to increase poverty, but reduces income inequality in countries where the central government represents a significant share of the economy.[18] Rodríguez-Pose and Ezcurra (2010) performed a cross-country analysis and concluded that decentralisation contributed to a reduction of regional inequality in high income countries, while it raised regional disparities in low and medium income countries.[19]

Keywords and Definitions

Block transfers	Unconditional transfers, which provide broad support in a specific area of subnational governments' performance
Conditional transfers	Transfers, which can be used by subnational governments under certain restrictions imposed by the upper level of government
Decentralisation	The transfer of authority and responsibility for public functions from the central government to subnational governments and/or the private sector

[17] For an analysis of a specific programme in Bangladesh, see Galasso and Ravallion (2000).

[18] Lindaman and Thurmaier (2002) found evidence that greater fiscal decentralisation improves the Human Development Index developed by the United Nations Development Programme. For an analysis using data of Ugandan households and public officials, see Livingston and Azfar (2010).

[19] For analysis of individual countries, see Qiao et al. (2008) for China, Faguet (2008) for Bolivia and Colombia, and Zakaria (2013) for Indonesia.

Externalities	Positive or negative effects, which are infeasible to charge, imposed by one economic agent on another without her consent
Intergovernmental fiscal transfers	Transfer of financial resources from an upper to a lower level of government
Public goods	Goods that have non-rival and non-excludable consumption
Unconditional transfers	Transfers, which may be used according to the preferences of the recipient subnational government

References

Abdelhak F, Chung J, Du J, Stevens V (2012) Measuring decentralization and the local public sector: a survey of current methodologies. IDG Working Paper No. 2012-01
Ahmad E, Brosio G (eds) (2006) Handbook of fiscal decentralization. Edward Elgar, Cheltenham
Ahmad E, Tanzi V (2002) Managing fiscal decentralization. Routledge, Oxford
Ahmad JK, Devarajan S, Khemani S, Shah A (2005) Decentralization and service delivery. The World Bank, Policy Research Working Paper No. 3603, Washington, DC
Akin O, Hutchinson P, Strumpf K (2005) Decentralisation and government provision of public goods: the public health sector in Uganda. J Dev Stud 41(8):1417–1443
Allers MA, Ishemoi LJ (2011) Do formulas reduce political influence on intergovernmental grants? Evidence from Tanzania. J Dev Stud 47(12):1781–1797
Asthana AN (2010) Decentralisation and supply efficiency: the case of rural water supply in Central India. J Dev Stu 39(4):148–159
Bahl R, Martinez-Vazquez JW (2002) State and local government choices in fiscal redistribution. Nat Tax J LV(4):723–742
Beddies S, De Soto H (2006) Albania: decentralization and water sector privatization. In:Coudouel A, Dani A, Paternostro S (eds) Poverty and social impact analysis of reforms. The World Bank/ UNDP, pp 293–338
Besley T, Coate S (2003) Centralized versus decentralized provision of local public goods. J Pub Econ 87(12):2611–2637
Bird RM, Smart M (2002) Intergovernmental fiscal transfers: international lessons for developing countries. World Dev 30(6):899–912
Boadway R, Shah A (2007) Intergovernmental fiscal transfers. Principles and practice. The World Bank, Public Sector Governance and Accountability Series, Washington, DC
Bucovetsky S, Smart M (2006) The efficiency consequences of local revenue equalization: tax competition and tax distortions. J Pub Econ Theory 8(1):119–144
Callan SJ, Thomas J (2001) Economies of scale and scope: a cost analysis of municipal solid waste services. Land Econ 77(4):548–560
Case A (2001) Election goals and income redistribution: recent evidence from Albania. Eur Econ Rev 45:405–423
Cox GW, McCubbins MD (1986) Electoral politics as a redistributive game. J Polit 48:370–389
Dasgupta S, Deichmann U, Meisner C, Wheeler D (2005) Where is the poverty–environment nexus? Evidence from Cambodia, Lao PDR, and Vietnam. World Dev 33(4):617–638
Diaz-Cayeros A, Magaloni B, Weingast BR (2007) Tragic brilliance: equilibrium party hegemony in Mexico. Working Paper, Hoover Institution, Stanford University

Dinar A, Kemper K, Blomquist W, Kurukulasuriya P (2007) Whitewater: decentralization of river basin water resource management. J Policy Model 29:851–867

Dixit A, Londregan J (1996) The determinant of success of special interests in redistributive politics. J Polit 58(4):1132–1155

Eaton K (2006) The downside of decentralization: armed clientelism in Colombia. Secur Stud 15 (4):533–562

Enikolopov R, Zhuravskaya E (2007) Decentralization and political institutions. J Pub Econ 91:2261–2290

Faguet J (2008) Decentralisation's effects on public investment: evidence and policy lessons from Bolivia and Colombia. J Dev Stud 44(8):1100–1121

Faguet J, Sánchez F (2008) Decentralization's effects on educational outcomes in Bolivia and Colombia. World Dev 36(7):1294–1316

Fedelino A, Ter-Minassian T (2010) Making fiscal decentralization work: cross country experiences, vol 271. International Monetary Fund Occasional Paper, Washington, DC

Foster V (2005) Ten years of water service reform in Latin America: toward an Anglo-French model. The World Bank, Water Supply and Sanitation Sector Board Discussion Paper No. 3, Washington, DC

Galasso E, Ravallion M (2000) Distributional outcomes of a decentralized welfare program. World Bank Policy Research Working Paper No. 2316, Washington DC

Geys B, Konrad KA (2010) Federalism and optimal allocation across levels of governance. In: Enderlein H, Wälti S, Zürn M (eds) Handbook on multi-level governance. Edward Elgar Publishing, Cheltenham, UK, pp 32–46

Gómez-Reino JL, Martinez-Vazquez J (2013) An international perspective on the determinants of local government fragmentation. In: Lago-Peñas S, Martinez-Vazquez J (eds) The challenge of local government size. Edward Elgar Publishing, Cheltenham, UK, pp 8–54

Habibi N, Huang C, Miranda D, Murillo V, Ranis G, Sarkar M (2003) Decentralization and human development in Argentina. J Hum Dev 4(1):73–101

Hecock RD (2006) Electoral competition, globalization, and subnational education spending in Mexico, 1999-2004. Am J Polit Sci 50(4):950–996

Hicken A, Simmons JW (2008) The personal vote and the efficiency of education spending. Am J Polit Sci 52(1):109–124

Inman RP, Rubinfeld DL (2012) Federal institutions and the democratic transition: learning from South Africa. J Law Econ Organ 28(4):783–817

International Monetary Fund (2013a) Government finance statistics. IMF, Washington, DC

International Monetary Fund (2013b) World economic outlook, October 2013: world economic and financial surveys. IMF, Washington, DC

Ivanyna M, Shah A (2014) How close is your government to its people? Worldwide indicators on localization and decentralization. Econ Open Access, Open Assess E J 8:2014–3

Khaleghian P (2004) Decentralization and public services: the case of immunization. Soc Sci Med 59:163–183

Khemani S (2007) Does delegation of fiscal policy to an independent agency make a difference? Evidence from intergovernmental transfers in India. J Dev Econ 82(2):464–484

Kurian M (2010) Institutions and economic development: a framework for understanding water services. In: Kurian M, McCarney P (eds) Peri-urban water and sanitation services: policy, planning and method. Springer, Dordreth, pp 1–26

Lal R (2008) Laws of sustainable soil management. Agron Sustain Dev 29:7–9

Lindaman K, Thurmaier K (2002) Beyond efficiency and economy: an examination of basic needs and fiscal decentralization. Econ Dev Cult Change 50(4):915–934

Lindbeck A, Weibull JW (1987) Balanced-budget redistribution as the outcome of political competition. Pub Choice 52:273–297

Livingston JA, Azfar O (2010) Federalist disciplines or local capture? An empirical analysis of decentralization in Uganda. The Selected Works of Jeffrey A. Livingston. http://works.bepress.com/jeffrey_livingston/10. Accessed 17 April 2014

Local Development International—LDI (2013) The role of decentralization/devolution in improving development outcomes at the local level: review of the literature and selected cases. Local Development International—LDI, New York
Lockwood B (2002) Distributive politics and the costs of centralization. Rev Econ Stud 69 (2):313–337
Lockwood B (2008) Voting, lobbying, and the decentralization theorem. Econ Polit 20 (3):416–431
Myerson R (2006) Federalism and incentives for success in democracy. Quart J Polit Sci 1:3–23
Oates WE (1999) An essay on fiscal federalism. J Econ Liter 37(3):1120–1149
Oates WE (2005) Toward a second-generation theory of fiscal federalism. Int Tax Pub Finan 12:349–373
OECD (2013) OECD fiscal decentralisation database. OECD, Paris
Qiao B, Martinez-Vazquez J, Xu Y (2008) The tradeoff between growth and equity in decentralization policy: China's experience. J Dev Econ 86:112–128
Rao G, Singh N (2005) Political economy of federalism in India. Oxford University Press, New Delhi
Robalino DA, Picazo OF, Voetberg A (2001) Does fiscal decentralization improve health outcomes? Evidence from a cross-country analysis. The World Bank Policy Research Working Paper No. 2565, Washington, DC
Rodríguez-Pose A, Ezcurra R (2010) Does decentralization matter for regional disparities? A cross-country analysis. J Econ Geogr 10(5):619–644
Rogoff K (1990) Equilibrium political budget cycles. Am Econ Rev 80:21–36
Santos BS (1998) Participatory budgeting in Porto Alegre, Brazil: toward a redistributive democracy. Polit Soc 26(4):461–510
Sepulveda CF, Martinez-Vazquez J (2011) The consequences of fiscal decentralization on poverty and income equality. Environ Plann C: Govern Policy 29:321–343
Shah A (2007) A practitioner's guide to intergovernmental fiscal transfers. In: Boadway R, Shah A (eds) Intergovernmental fiscal transfers. Principles and practice. The World Bank, Washington, DC, pp 1–53
Shah A (2008) A global dialogue on federalism. In: Shah A (ed) The practice of fiscal federalism: comparative perspectives. McGill-Queen's University Press, Montreal
Shelton C (2007) The size and composition of government expenditure. J Pub Econ 91 (11–12):2230–2260
Simatupang RR (2009) Evaluation of decentralization outcomes in Indonesia: analysis of health and education sectors. Economics Dissertations, Paper 58, Georgia State University
Smoke P (2006) Fiscal decentralization policy in developing countries: bridging theory and reality. In: Banguara Y, Larbi G (eds) Public sector reform in developing countries. Palgrave Macmillan, London, pp 195–227
Stepan A (2004) Toward a new comparative analysis of democracy and federalism: demos constraining and demos enabling federations. In: Gibson EL (ed) Federalism and democracy in Latin America. Johns Hopkins University Press, Baltimore, pp 29–34
Tanzi V (2005) The economic role of the state in 21st century. Cato J 25(3):617–638
Tanzi V, Schuknecht L (2000) Public spending in the 20th century. Cambridge University Press, Cambridge, UK
The World Bank (2013) Fiscal decentralization indicators. The World Bank, Washington, DC
United Cities and Local Governments (2010) Local government finance: the challenges of the 21st century. Second Global Report on Decentralization and Local Democracy
Vásquez WF, Franceschi D (2013) System reliability and water service decentralization: investigating household preferences in Nicaragua. Water Resour Manag 27:4913–4926
Weingast BR (2014) Second generation fiscal federalism: political aspects of decentralization and economic development. World Dev 53:14–25
Wilder M, Lankao PR (2006) Paradoxes of decentralization: water reform and social implications in Mexico. World Dev 34(11):1977–1995
Zakaria S (2013) The impact of fiscal decentralization toward regional inequalities in eastern region of Indonesia. J Econ Sustain Dev 4(10):74–84

Chapter 2
Public Budgets: Governance Structures, Norms, and Organizational Practices

Abstract This section focuses on public budgets and, subnational governments' expenditure and revenue choices. It begins with a general picture of public budgeting and of global trends in budgeting, such as performance budgets, the adoption of medium-term expenditure frameworks, and the creation of independent fiscal institutions. It then addresses the subnational governments' expenditure responsibilities. It explains the main drivers of subnational governments' expenditure needs and costs, approaches to estimating costs, the concepts of zero-based budgeting and participatory budgeting, and the increase of private participation in service delivery. Finally, issues regarding subnational governments' revenue options, such as taxes, user fees, and debt are discussed.

Keywords Budget · Fiscal autonomy · Expenditure responsabilities · Participatory budgeting · Performance budget

This section provides the basic tools for understanding public budgeting and subnational governments' expenditure and revenue choices.[1] It includes four subsections. The first lays out an overall framework for budgeting. It explains the notion of budget, its components and structure, and the budget cycle. Recent trends in budgeting, such as performance budgets, the adoption of medium-term expenditure frameworks, and the creation of independent fiscal institutions are also addressed. The second subsection analyzes patterns of expenditure responsibilities of subnational governments across countries. It discusses the main drivers of subnational governments' expenditure needs and costs and presents approaches to estimating costs. It explains the concepts of zero-based budgeting and participatory budgeting and examines the trend toward decentralization and increased private participation in public service delivery. The third subsection deals with subnational governments' fiscal autonomy. An overview of the main sources of subnational governments' revenues across countries is presented. The main questions involving the taxes levied and the user fees charged by subnational governments are debated. Finally, the last subsection is dedicated to debt. It discusses the common pool

[1] See Shah (2007a) for a comprehensive book on local budgeting.

L.G. Veiga et al., *Intergovernmental Fiscal Relations*,
SpringerBriefs in Environmental Science, DOI 10.1007/978-3-319-06296-9_2

problem and other political economy issues that contribute to subnational governments' fiscal indiscipline in several countries. It also examines whether the establishment of fiscal rules promotes sound fiscal policy.

1 Budget Basics

Any organization, private or public, has to prepare a budget to manage its resources and activities properly, in order to accomplish its goals. While in the private sector most organizations have profit maximization as their key goal, the public sector's main objective is to improve the welfare of the population. Budgets are an important tool to discipline governments because the services they provide (public goods, correction of externalities, and other market failures) are not subject to market competition as are those supplied by private businesses. At the subnational level, budgetary institutions should serve to allocate the community's resources to their most preferred activities, e.g., those that best satisfy their needs.

1.1 What Is a Budget?

A budget is a document or a collection of documents comprising a detailed description of the expected revenues and expenditures of a given institution, associated with the activities that are planned for achieving specific purposes or goals, within a given period.

In the case of subnational governments, the local budget is intended to link the community needs with the amount of resources necessary to satisfy them, serving as a guide for local financial management and as a tool that can be used by citizens to evaluate subnational governments performance and fiscal discipline. Therefore, budgeting is fundamental for the planning, control, and evaluation process of governments.

In the public sector, budgets have three broad functions: economic, political, and legal. Economic, because they are an exercise of planning, controlling, and administering activities, intended to balance revenues and expenditures, and to allocate available resources efficiently in order to maximize social welfare. The budget establishes which activities will be undertaken (and therefore, the type, quantity, and quality of services provided to citizens) and how resources will be obtained and allocated. Budgets help officials to check if the flows of revenues and expenditures during the fiscal period materialize as planned, and if operational adjustments are necessary. At the end of the fiscal period, comparisons of budgets with accounting final reports allow an evaluation of whether the flows of revenues and expenditures were in accordance with expectations or not. If linked to the subnational government objectives, budgets allow for an assessment of its efficiency.

The political function of the budget results from the fact that budget proposals offered by the executive body of subnational governments have to be approved by committees that are usually elected. Furthermore, during the implementation of the budget and at the end of the fiscal period, officials have to present reports that require approval by legislative bodies. Finally, since budgets are regulated by laws, rules and regulations and, in some cases, they have a legal status, they carry out a legal function. They establish limits to managerial decisions and actions of subnational governments, and officials violating them may be subject to penalties.

There are several types of budgets. Operating (or current) budgets focus on ongoing operations while capital budgets are concerned with new long-term assets (such as land, infrastructures and equipment). In some countries, the law requires that subnational governments have balanced operating budgets. As for capital budgets, since they concern the acquisition and construction of assets that generate benefits over several years, inter-period equity suggests that they should not be financed by taxes in a single year, but rather by debt that is repaid over a period that extends farther into the future. Additionally, there are special fund budgets, which cover programs (e.g., highways) funded by specific revenue sources (e.g., sales taxes on gasoline).

The way the budget is prepared and presented, in terms of its components and structure, influences the allocation of resources and the government's capacity to monitor and control its activities. Therefore, considerable attention has been dedicated to establishing the desirable form and content of budgets.

To allow for comparisons over time and across governments, it is important to establish standard classifications for reporting revenues and expenditures, and rules for the structure and elements of budgets. There is usually more concern about reporting expenditures than revenues because governments, particularly at the subnational level, have power that is more discretionary over the former than the latter. Typically, at least three classifications of expenses are adopted: functional, organizational, and economic. For revenues, the economic classification is the most common. Classification systems vary from country to country. International organizations, such as the International Monetary Fund (IMF) or the Organization for Economic Cooperation and Development (OECD), also have standard classification systems that allow for cross-country comparisons.[2] The IMF economic classification of expenses includes eight categories (compensation of employees, use of goods and services, consumption of fixed capital, interest, subsidies, grants, social benefits, and other expenses) divided into subcategories. The functional classification has 10 components (general public services; defense; public order and safety; economic affairs; environmental protection; housing and community amenities; health; recreation, culture, and religion; education; and social protection) also divided in subcomponents. Revenues are divided into taxes, social contributions, grants, and other revenues.

[2] For a detailed description see the IMF's Government Finance Statistics Manual 2001.

Local governments can only budget effectively if they have the autonomy to determine expenditures and revenues and discretion for the use of funds, which depends on the degree of political, administrative and fiscal decentralization (Mullins 2007). Budgeting in developing and transition countries presents additional challenges. On the revenue side, many local governments face difficulties in increasing their funding resources. On the expenditure side, budgets mostly cover operating costs. To ensure the efficiency and sustainability of local services, namely the management of environmental resources, good governance practices and clear accountability are fundamental.

1.2 Budget Cycle

Public budgeting varies considerably across the world. It depends on the country's legal framework, as well as on the organizational structure of government.[3] The budgeting process involves interaction among numerous participants, from citizens to firms, and includes officials from various levels of government. Although budgets are usually set for a fiscal year, the budgeting process extends for a considerably longer period and is best understood as a cycle with overlapping phases, as shown in Fig. 1.

Four key general phases of the budgeting system can be identified.

(a) *Formulation and submission*: in most cases, the executive body of government prepares and submits the budget for approval. The initial phase of the budget involves flows of information on spending requests and estimates of the resources available, as well as decisions on how scarce resources should be allocated among numerous public demands to attain the community's goals and objectives. The budget formulation is both a technical and a political process. Once finished, the budget proposal is submitted to the legislative body.

(b) *Debate and adoption*: the legislative branch analyzes, debates, and proposes changes to the budget proposal. During this stage, the budget proposal is also frequently disclosed to the wider public. After being approved by the legislative body, the amended budget is officially adopted by the legislature and is put into effect.

(c) *Execution*: the executive body implements the budget by operationalising plans, collecting revenues and spending money. Frequently, several control mechanisms are adopted during the budget's implementation, including cash management, audit-systems, appropriation and allotment rules, and transfers of authority. Accounting and reporting procedures are implemented to guarantee that revenues and expenditures are continuously monitored and stay on track with the amounts that have been estimated and authorized. The executive

[3] See Guess and Leloup (2010) for a comparison of public budget systems around the globe.

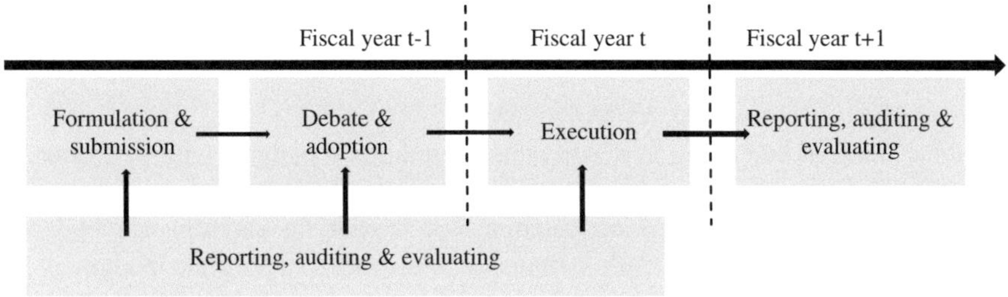

Fig. 1 The budgeting cycle

branch has discretionary power to implement small budget adjustments, but substantial changes often require the approval by a higher level of government or by the legislative branch.

(d) *Reporting, auditing and evaluating*: Public officials should be accountable to citizens for how they raise public monies and how they spend them. To guarantee accountability and keep stakeholders informed, frequent evaluation and reporting are critical. Although most citizens lack the time or knowledge to read governments' financial reports, they are often exposed to debates during electoral periods, which are used by opposition parties to attack the incumbent. Before, during and after the budget execution, several types of audits are implemented. Fiscal audits are intended to verify the accuracy of expenditure and revenue records, to determine if actual financial results are in accordance with the legally adopted budget, and to assess whether financial transactions comply with finance-related laws, rules and regulations. Operational and management audits review how specifically programs are carried out and evaluate the efficiency and effectiveness of management. Performance audits verify if the outputs and outcomes are in accordance with the benchmarks previously defined. Audits can be internal or independent. The increase of governments' deficits and debts during the recent global crisis, has led to the creation of independent fiscal institutions in a growing number of countries.[4] Common functions of these councils include the analysis of fiscal policy and budget proposals, the production or endorsement of macroeconomic forecasts, monitoring compliance with fiscal rules, and analysis of long-term fiscal sustainability. Essentially, these entities were created to increase fiscal policy transparency and soundness.

[4] In the European Union new regulations require member states to have independent fiscal institutions. For discussions on the role of independent fiscal agencies see Hagemann (2011), Kopits (2011), and Hemming and Joyce (2013).

1.3 Performance Budgets and Medium-Term Expenditure Frameworks

A global trend in budgeting is to supplement traditional budgets with performance budgets. These move the attention away from control of expenses and revenues toward responsibility for service outcomes or results. In performance budgets, expenditures are associated with outputs or outcomes. A program budget is an example of a performance budget, where resources and results are associated with specific programs. While subnational governments may lack technical skills in statistical monitoring and may be unable to produce a sophisticated evaluation of results, they can make up for that with their greater proximity to the citizenry being served. By paying attention to public response and integrating public input into the monitoring process, a results or outcome orientation emerges. Local services' results and outcomes are easier to monitor than those of central services are.

Budgets can be produced for a single fiscal year or for several years. Another important recent change in budgeting and public financial management is the widespread adoption of medium-term expenditure frameworks (MTEF). According to World Bank (2013), by the end of 2008, more than 75 % of all countries had adopted a MTEF. Most public programs require resources and generate benefits over several years. Under these circumstances, single-year budgets may not provide adequate information. MTEFs ensure a multiyear commitment of resources to policies and are, therefore, important for expenditure prioritization and for fostering government performance over the medium term. To implement them, governments have to consider possible trade-offs between short- and medium-term goals and equity issues over time. International organizations, such as the World Bank, included MTEF in their standard advice on budget reforms to low and middle-income countries, contributing to their spread around the world. World Bank (2013) provides strong empirical evidence, based on case studies and econometric analysis, that MTEFs enhance fiscal discipline and allocative efficiency.

2 Expenditure Responsibilities of Subnational Governments

Subnational governments vary in their responsibilities across countries. As was studied in Sect. 1, they play an increasing role in public services delivery. Their expenditures represent an important share of total public expenditure in sectors such as environmental protection, housing, education, and health.

2.1 Defining Expenditure Needs and Costs

Expenditure needs of subnational governments are higher in more decentralized countries and where they are assigned a broader range of public services. In some countries, the assignment of responsibilities across levels of government is defined by the constitution (e.g., Switzerland and South Africa), in others by statute at the national or provincial level. Frequently, the set of functions assigned to subnational governments depends on the size (population) of the jurisdiction with capitals of metropolitan areas being responsible for more functions. Services such as sanitation, garbage removal, street repair and cleaning, fire protection and recreation facilities are the responsibility of subnational governments in almost all countries. The assignment of other functions, namely education and health care, which are more costly, varies across countries. According to Shah and Shah (2006), subnational governments in developing countries have limited autonomy in expenditure decisions and they are typically allowed to discharge a number of functions, which are mandated from above.

As can be seen in Fig. 2, most subnational governments devote the largest part of their budget to education, followed by economic affairs, general public services and social protection. Non-OECD countries included in the graph spend less in environmental protection and recreation, culture and religion than OECD countries do, but they spend more on economic affairs and housing and community amenities.

Expenditures are determined by not only the degree of decentralization, but also by the functions assigned to subnational governments and the needs of the population. They are also determined by supply side factors influencing the costs of

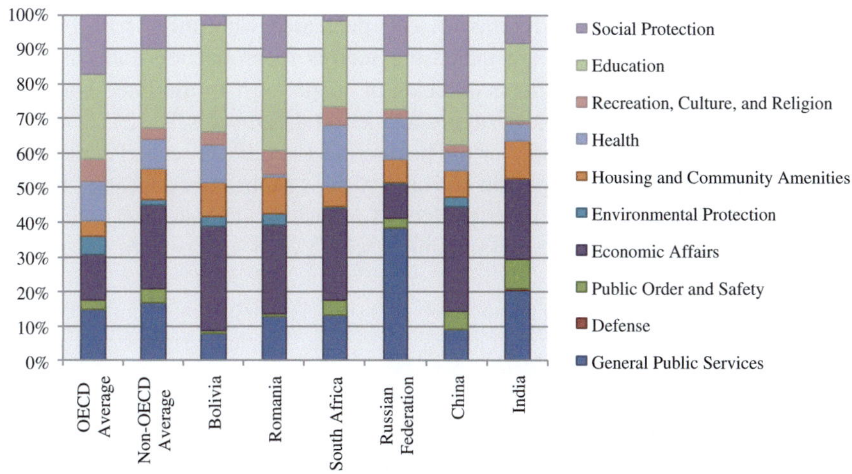

Fig. 2 Expenditure composition for subnational governments. *Source* Own calculations based on the IMF's Government Financial Statistics. *Note* For each country, the latest data available was used. The non-OECD average refers to the countries considered in the figure

public service provision, such as the level and quality of services provided and inefficiencies in provision resulting from mismanagement.

The costs incurred in delivering a given level of a public service vary across subnational governments. The main factors influencing costs are the quantity and composition of inputs needed given the available technology, the prices of inputs, physical features of the region (altitude, temperature, and other environmental factors), and the social and demographic characteristics of the population.

Labor costs are frequently an important part of total costs incurred by subnational governments. They depend on the wage paid and the number of workers hired. Frequently, public employees receive the same wages regardless of the subnational government hiring them. However, in some countries subnational governments can use higher compensation to attract workers with better qualifications or stimulate them to move to remote or high criminality areas. Non-labor costs are associated with inputs such as raw materials, energy, land, and equipment. Capital expenditures are a special type of non-labor local expenditure. They relate to the expenditure to acquire capital assets. The physical characteristics of the place where public goods and services are delivered also influence costs. For example, roads, sewage, water and energy are easier to provide in flat areas than in areas with high altimetric amplitude. Weather conditions, such as temperature and precipitation are also relevant issues. Additionally, economies of scale, density and scope may also influence costs. Several empirical studies suggest that many public services have U-shaped average costs. For public services with high-fixed costs and relatively low operating costs, costs per unit of output decline as the scale of production increases (economies of scale). However, after a certain level of production, economies of scale are exhausted and average costs increase with scale. Nauges and van den Berg (2008) analyzed water and sewerage utilities in Brazil, Moldava, Romania and Vietnam. They found evidence of economies of scale in the last three countries; however, returns to scale decreased with utility size. Finally, their results reveal that there are increasing returns to production and customer density in all countries, and that economies of scope exist in the three countries where utilities provide both water supply and sewerage services.[5]

The socio-demographic profile of the population also affects costs. People from different age groups demand specific services, such as elementary education for the youngest and facilities for the elderly. Needs for basic health care and social protection are larger in low-income communities with a high incidence of poverty.

2.2 Approaches to Estimating Costs

A budget is a financial plan that is oriented toward the future. Accurately forecasting future expenditures and resource needs is critical to good local financial

[5] For another study on returns to scale, see Reingewertz (2012).

management and to the achievement of established goals. According to Schroeder (2007), four general forecasting techniques can be used for expenditures and revenues. The first is judgemental forecasting, that is, to rely on local experts to generate forecasts. This approach may be necessary when data on the public sector is limited, but dangers are inherent when the quality of the forecast depends entirely on the knowledge of the experts. The second technique is to apply time series analysis to project the next year's flows based on what has happened in the recent past. This method implicitly assumes that the variables that influenced expenditures and revenues in the past will continue to influence them in the future in a similar manner. The third method consists of deterministic techniques. For example, to predict the labor expenditure of a local government, a forecaster calculates the product of the number of employees and the average annual wages and salaries, making assumptions about the value each of these variables will take. This technique does not require the assumption that revenues or expenditures will rise or fall as they did in the past, but explicit assumptions about the factors determining the variables being forecasted must be made,[6] and these assumptions may turn out to be wrong. Finally, there are statistical (or econometric) forecasting models that while the most sophisticated might require more data.

Let's now focus on estimating costs. Some costs are relatively easy to estimate, either because they are fixed by legislation (such as salaries) or ruled by contracts (e.g., equipment or debt payment). However, others are more discretionary or unexpected and, therefore, much less predictable (e.g., equipment repair or recovery from natural disasters). Reschovsky (2007) proposes three approaches to estimating the costs of local public services. When good data on public outcomes is available, it may be possible to estimate a cost function. Cost functions are statistical relationships between expenditures on a specific public service, outcomes of the public service, prices of the inputs necessary to produce the public service, environmental factors that may influence the relationship between inputs and output, and characteristics of local governments that may influence spending. Estimation of cost functions can be very demanding in terms of data. For services such as water and energy distribution or garbage collection, output is relatively easy to measure. Regarding water, the number of households having access to potable water or the number of liters consumed can be used. However, for other services such as police protection, education, or health care, measuring outputs is more difficult. An alternative method is to estimate an expenditure equation that relates per capita expenditure on a specific output with input prices, environmental factors that might influence the relationship between inputs and outputs, and variables that determine preferences for the public good. Finally, when gathering data is problematic, it is possible to rely on a panel of experts to identify the minimum cost of providing a specific public service.

[6] For example, regarding future salaries or wages, one possible assumption is that they will increase at the same rate as inflation. Expected values for the amount of labor to be employed should depend on the amount of output to be produced.

2.3 Zero-Based Budgeting and Participatory Budgeting

Mapping expenditures to the perceived preferences of citizens is a complex process that is time consuming and involves high costs. Given the limited period of time within which the budget is prepared, and in order to avoid conflicts over budgetary options, in traditional budgeting only marginal changes to previous base budgets are introduced. Most programs are rolled over from one year to the other, thus avoiding the difficult process of negotiating their discontinuance with stakeholders. Since the 1970s, it has become clear that incremental budgeting leads to an upward bias in spending that can contribute to the growth of government and to higher deficits. In developing countries, where resources are very scarce relative to the development needs, the negative effects of incremental budgeting practices are even higher. Therefore, efforts have been made to overcome these shortcomings.

In the late 1970s, zero-based budgeting was introduced in the US. As the name suggests, at the beginning of each budget cycle the starting point is zero so that the continuation of programs is not guaranteed. Each spending agency has to justify every item in its request for funding. This approach allows the discontinuation of programs that are no longer necessary so that resources are used based on needs and not on decisions taken in the past. In practice, zero-based budgeting is difficult to implement due to the high costs involved. First, it is time consuming to justify each item included in the budget and human resources to perform the needed evaluations are often lacking. Second, some pubic programs are not easily negotiable because they are regulated by legislation and involve multi-year contracts. Third, zero-based budgeting does not guarantee that low priority programs will be rejected because it does not take into account the political pressures and institutional features that drive budgetary decisions.

Another effort to make budgeting more rational is participatory budgeting.[7] It relates to the direct participation of citizens in deliberations and negotiations over public spending and resource allocation. Traditionally, citizens' participation in subnational governments has been political. Through campaigns, voting and lobbying, citizens try to influence public representatives and officials. Some argue that citizens' views should be incorporated earlier, starting when budget proposals are being defined. In a very simplified way, participatory budgeting can be described by the following sequence of steps: residents propose and discuss spending ideas, budget delegates prepare proposals based on these ideas, residents vote on the proposals and the government implements the proposal selected. This enables taxpayers to work with the government to make the budget decisions on how to spend part of a public budget. It engages people in government by ensuring that communities are well-represented in the different phases of the budget process

[7] For an overview of the principles underlying participatory budgeting, regional surveys and country case studies, see Shah (2007b). For web information consult the site of the Participatory Budgeting Project (http://www.participatorybudgeting.org/), a non-profit organzation that helps communities decide how to spend public money, primarily in the US and Canada.

(planning, development, approval). It gives an opportunity to marginalized and excluded groups of the population to influence subnational governments' policies in accordance with their interests.

Participatory budgeting was first introduced in 1989, in the city of Porto Alegre in Brazil and it significantly improved facilities in the town, particularly those that benefited the poor.[8] For example, sewer and water connections increased from 75 % of total households in 1988 to 98 % in 1997. Given the noticeable positive impact of PB in citizens' lives, it has spread across the world. According to Mikesell (2007), participatory budgeting has been successfully implemented by local governments in several countries, such as Brazil, Canada, China, the Dominican Republic, Ecuador, India, Indonesia, the Philippines, Serbia, South Africa, Sri Lanka, Tanzania, the United Kingdom, and Uruguay. There is broad variation in the design of PB programs, as they are conditioned by the local social, political and economic environment.

The more open and inclusive policymaking processes provided by participatory budgeting help to ensure that policies are better informed and will better match citizens' needs. Well-implemented participatory budgeting improves subnational governments' performance and enhances the quality of democracy. However, as mentioned by Shah (2007b), participatory budgeting also has significant dangers that need to be avoided. When participatory budgeting functions poorly, it can be captured by interest groups, lead to cynicism about democracy, and diminish long-term planning, as many participants are only interested in securing short or medium-term public works projects.

2.4 Private Participation in Service Delivery

To increase the match between citizens' needs and service delivery, several countries have decided to end public monopolies in areas including sewage disposal, and water and electricity distribution. Private sector participation (World Bank 2004) in these areas has been permitted. When possible, the emergence of private competition where citizens can choose the provider based on price and output quality gives citizens a voice and may ensure a better match between demand and supply. However, for infrastructure services requiring high fixed costs and involving strong economies of scale (i.e., water and electricity distribution) or involving network externalities, a single distributer provider may be technically more efficient. Additionally, government intervention may be justified for equity reasons, to guarantee minimum equitable access to the services. In this case, two arrangements are possible. The first is to have a public sector provider. Most of

[8] For an analysis of the Brazilian case, see the World Bank report available at http://www-wds.worldbank.org/external/default/WDSContentServer/WDSP/IB/2009/11/03/000333037_20091103015746/Rendered/PDF/514180WP0BR0Bu10Box342027B01PUBLIC1.pdf.

these providers belong to subnational governments but in some cases to central governments. When the central government is responsible for regulating and delivering the service it has little incentive to hold itself accountable. Devolution of responsibilities to subnational governments creates opportunities to benchmark performance and provide fiscal incentives to stimulate efficient service delivery. However, according to Foster (2005),[9] devolution may also lead to losses of scale economies, difficulties of attracting private investment into a highly fragmented sector, and conflicts that arise when a central government imposes regulation in a sector that is legally under municipal control. The second arrangement is for the government to create incentives for private provision, regulate the service and monitor its behavior. Private participation may increase management expertise and private investment, and lead to the introduction of incentives for high quality service delivery. Like decentralization, private provision may enhance accountability by separating providers from policymakers.[10] The success of reforms depends crucially on how the legal and policy frameworks are designed and on the accountability of subnational authorities (Kurian 2010). In recent years, output-based approaches that tie public disbursement of funds to specific service or output deliveries are being increasingly used as a tool to increase access to services.

In industrialized countries, the trend toward public ownership and private provision occurred in the Reagan-Thatcher era. In France and the USA, water assets have been assigned to local governments. To take advantage of economies of scale, clusters of political jurisdictions were created to delegate water services to private or semi-public companies. In Bolivia, Colombia, and Brazil, inter-jurisdictional agreements also exist in the water services. In England, the local water system was consolidated into regional bodies after World War II and they were privatized in the 1980s. A centralized agency relying on incentive-based instruments regulates the market. Since the 1990s, private participation in water, sanitation and electricity has grown significantly (World Bank 2004).

3 Local Governments' Fiscal Autonomy

A wide portfolio of revenue sources improves government's risk sharing, helps to handle economic fluctuations and reduces the impact of unexpected events in revenues. Therefore, it is important for subnational governments to diversify their sources of revenue. State and local governments receive revenue from fiscal transfers, a variety of taxes, charges and fees, and borrowing. Fiscal transfers

---

[9] Foster (2005) analyzed water service reform in Latin America, during the 1990s, where most countries moved from unregulated centralised public provision of water to regulated decentralised public provision.

[10] This topic will be developed further in the next section.

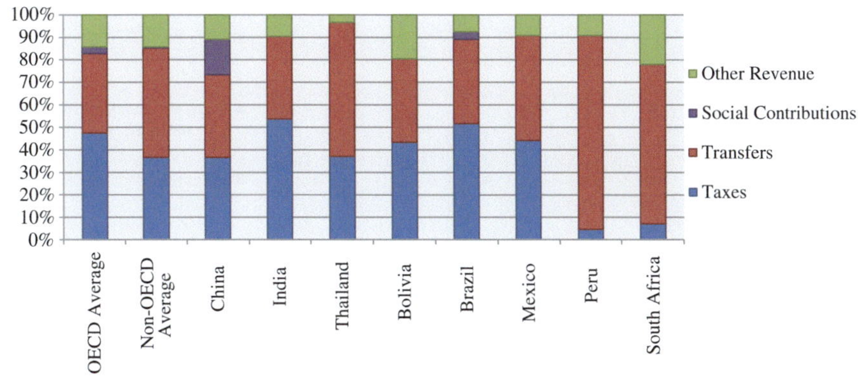

Fig. 3 Composition of revenues for subnational governments. *Source* Own calculations based on the IMF's Government Financial Statistics. *Note* For each country, the latest data available was used. The non-OECD average refers to the countries considered in the figure

include intergovernmental transfers within the country, as well as transfers from foreign governments and international organizations.

In terms of budgeting, revenues are relatively hard to predict accurately because they depend on exogenous factors such as economic performance at the local and national levels (tax revenues) and citizen's preferences (fines and fees). Revenues that are contractually established, promised in advance by other governments or institutions, or set by law like property tax, are easier to estimate. Revenue estimation errors may lead to resource misallocation and to the accumulation of debt.

As can be seen in Fig. 3, the large majority of subnational governments' revenue comes from taxes and intergovernmental fiscal transfers. In OECD countries, almost 50 % of revenue is collected through taxes, while in developing countries the largest part comes from intergovernmental fiscal transfers. For example, in Peru and South Africa, subnational governments are extremely dependent on transfers. In these countries, tax decentralization has not kept pace with political and expenditure decentralization (recall that subnational governments' expenditures represent 20 % of total public expenditure in Peru and 50 % in South Africa—Table 1). This separation of taxing and spending decisions undermines accountability to local citizens, since it reduces subnational politicians' need to explain their decisions to their electorates. However, in other countries such as Brazil and India, subnational governments rely more on self-financing, which is important for strengthening governance, efficiency and accountability. In China, around 16 % of subnational revenues are obtained from social security contributions, which is the largest share in the sample. This is because Chinese local governments have responsibilities in the social security area, namely in terms of pensions and unemployment subsidies.[11] Items included in Other Revenue represent more than

[11] For a review of subnational taxation in Brazil, Russia, India, China and Nigeria see Bird (2012).

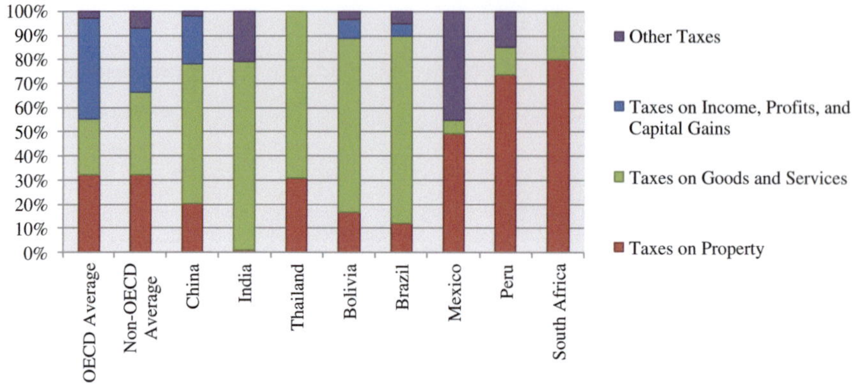

Fig. 4 Composition of subnational governments' tax revenues. *Source* Own calculations based on the IMF's Government Financial Statistics. *Note* For each country, the latest data available was used. The non-OECD average refers to the countries considered in the figure

10 % of subnational governments' revenues. They include, among others, property income, sales of goods and services, and fines, penalties and forfeits.

Figure 4 shows that the composition of subnational governments' tax revenues varies substantially across countries. For OECD countries, revenues from taxes on income, profits and capital gains are the largest, followed by taxes on property. For non-OECD countries, taxes on goods and services and property taxes have the biggest share. Mexico, Peru and South Africa rely mostly on property taxes, while China, India, Thailand, Bolivia, and Brazil rely on taxes on goods and services.

3.1 The Property Tax

Taxes on property account for more than 30 % of tax revenues of subnational governments.[12] They are generally taxes on land and capital (buildings and equipment). An important advantage of the property tax is that it generates a certain, and therefore predictable income. However, taxes on property also have disadvantages. First, they are frequently regressive.[13] Although only landlords pay it, when property taxes increase rents also tend to go up because landowners transfer the burden of the tax to the tenants. Additionally, increases in property taxes are particularly painful for those with fixed incomes, such as retired people

[12] For an analysis of the property tax reform in Vietnam see McCluskey and Trinh (2013), for Kenya see Kelly (2002), and for Latin America see Sepulveda and Martinez-Vazquez (2011). McCluskey et al. (2012) provide a broader discussion on property tax.

[13] A regressive tax imposes a burden that is a greater fraction of income for lower-income individuals than higher-income persons. Taxes can also be progressive or proportional.

who may be forced to sell their homes. Second, property taxes are hard to administer and somewhat arbitrary. Unlike other taxes, subnational governments determine both the tax rate and the tax base of property taxes. The tax liability is the product of the tax rate and the property assessed value by the jurisdiction, most commonly the market value. However, when property is not sold frequently its value has to be estimated; otherwise, it may become increasingly under-assessed. Regardless of the practices and procedures used to perform the estimations, they are always subject to error.

When subnational governments rely extensively on property taxes, differences in the value of property across jurisdictions generate differences in their capacities to provide public goods and services. In wealthy communities, or in places with capital-intensive industries, or with a high construction rate it is easier for subnational governments to provide high levels of public services with low tax rates than in areas where the value of property is low.

3.2 Other Tax Revenue

In several countries, subnational governments are authorized to levy taxes on goods and services. In countries like India, Thailand and Brazil, they constitute a very important source of revenue for subnational governments. When applied to all merchandise, these taxes are clearly regressive. They place a proportionately higher burden on low-income people because they spend a higher proportion of their incomes than rich people do. To reduce its regressivity, most countries exempt certain goods and services (e.g., food for home consumption, drugs, electric and gas utilities) from taxation. Taxes on goods and services are usually easier to administer and govern than property taxes, but they may lead to strategic competition among subnational governments and it is difficult to collect taxes on purchases made by residents in other jurisdictions. Subnational governments may also benefit from miscellaneous other taxes on goods and services, such as taxes on tobacco products, alcoholic beverages, motor fuels, and on the extraction of natural resources. Severance taxes on natural resources help to pay for environmental damages resulting from extraction, and keep other taxes low. However, when the jurisdiction becomes highly dependent on them, a fall in the price of the natural resource can exert severe damage to the local economy.

Taxes on income, profits and capital gains are also important sources of revenue for subnational governments, particularly in OECD countries. They can be used strategically by subnational governments to attract population and businesses.

3.3 User Fees

User fees are classified as Other Revenue in the IMF's Government Financial Statistics classification of revenue (see Fig. 3). User fees include prices charged by subnational governments on goods and services they provide, license taxes and fees required to undertake some activity. Common examples are water and sewer charges, tolls on roads and bridges, bus charges, and parking fees.

Several advantages are associated with user fees. First, only those who use the good or service are asked to pay (benefit principle). Second, they allow for a more direct relationship between the services and their costs than with tax financing. Third, user fees can also be used to moderate consumption. They can be increased to reduce access to overcrowded facilities, or assume different values during the day of the week or month in order to smooth consumption and promote an efficient utilization of the good or service. Finally, persons who do not live in the jurisdiction but benefit from public services must also pay the price. However, user fees also have disadvantages. First, they fail the ability to pay principle, since poorer persons are those more in need for public services but they may not have enough income to support them. A second potential problem is that they may generate high administrative costs to the subnational government (e.g., measuring use and collecting the fee) and compliance costs to the users (e.g., time costs).

User fees are frequently used to finance water, sewer and garbage collection services. In the case of water, the fee usually comprehends a connection charge, a capital and distribution charge, and a water-supply charge (Fisher 2007). For sewer services, the rationale is similar. Because metering sewer discharges is not common, except for industry, the fee frequently varies according to the amount of water consumed. Fees for garbage collection should depend on the amount and type of garbage made by the household, in order to connect costs with financing and generate incentives to avoid refuse through recycling. However, it is not easy to measure the amount of refuse, so in practice garbage collection services are frequently financed through general taxes or a fixed fee per month.

In several developing countries, governments do not charge consumers for water and sanitation services because they are considered unable to pay for the services. However, some studies claim that several consumers are willing to pay higher prices to firms offering services that are more reliable (World Bank 2004). According to Kurian and Ardakanian (2015), this contradiction can be explained by the fact that: (a) subnational governments are highly dependent on intergovernmental fiscal transfers that finance the construction of technical options that are highly costly and involve operating costs that cannot be supported from local revenues; and (b) it is difficult to identify the individuals who in fact cannot afford the services to give them infrastructure connection subsidies, so nobody ends up paying. Furthermore, it is frequently difficult to identify households that are benefitting from the service but not paying, the staff managing the service frequently lacks incentives to increase the number of clients, which is decisive for per capita charges reductions, and politicians use grandiose infrastructure projects to woo the electorate.

4 Debt

Although constrained by law in most countries, subnational governments can issue debt to finance their activities. Borrowing is particularly relevant to finance long-life capital infrastructure when it is recognized that transfers from upper levels of government are insufficient to satisfy the needs of the population. Because such investments generate benefits over the long run, intergenerational equity recommends that the associated costs be spread over the asset's useful life period. The main approaches adopted by subnational governments to fund infrastructure are: direct access to financial markets; the creation of financial institutions to serve as intermediaries between local governments and lenders (by pooling risk and having a better knowledge of how financial markets operate, they facilitate access to loans by local governments); borrowing from development banks and other financial institutions; and engaging in public-private partnerships.[14]

The capacity and prudence of debt financing by subnational governments depends on the financial market situation, the creditworthiness of the government, and the type of expenditure to be financed. When subnational governments borrow, they must predict the future expenditures needed (debt service) to pay back the loan and the interest on it. Therefore, debt is a commitment of the government's future revenues. Thus, it must be issued and managed carefully, in order to avoid future problems of fiscal unsustainability and difficulties in accessing the debt market.

The interest rate paid on loans is a key variable that is determined by the availability of funds in the financial market and by the financial situation of the local government. Borrowing is self-limiting because when lenders have doubts about the financial situation of a government, they will demand higher interest rates on the loan, which in the extreme case may exclude the government from the debt market. An indicator that is frequently used to assess the creditworthiness of local governments is the ratio of net operating surplus to anticipated new debt service payments. To ensure financial sustainability, subnational governments should establish a debt policy that limits the total amount of debt and debt service, establishes parameters for the issuance of new debt, and informs citizens and the financial community of the local government's goal of developing sound fiscal policy.[15] Mikesell establishes the following principles for responsible local debt-management.

- "Long-term debt should be used only to acquire capital infrastructure that has a useful life longer than the maturity of the debt.
- Debt should not be refinanced to extend its maturity (although it may be refinanced to take advantage of lower interest rates).

[14] For details see Alam (2010). See OECD (2012) for principles for public governance of public-private partnerships.

[15] For an example of a simple local government debt policy that can be adapted as needed, see Kaganova (2011).

- Short-lived capital assets should be financed with current revenue rather than by borrowing.
- Local governments should integrate capital asset planning and financial planning to ensure the affordability of long-term infrastructure programs.
- Debt should be issued on a competitive basis.
- Localities should practice complete, comprehensive, and clear debt reporting" (Mikesell 2007, p. 34).

However, lack of fiscal discipline by subnational governments is a problem in several countries, particularly in developing ones. Below we present some explanations for the problem and discuss whether the establishment of fiscal rules promotes sound fiscal policy.

4.1 The Common Pool Problem and Other Political Economy Issues

One of the first explanations for fiscal indiscipline by subnational governments is the common pool problem. When subnational governments are highly dependent on transfers from higher levels of government, they tend to overspend and to accumulate budget deficits. Because of lack of coordination, local decision makers take full credit for additional spending in their jurisdiction and the benefits it brings to the residents, but fail to internalize fully the costs that all taxpayers must bear. The problem is aggravated if subnational governments expect to be bailed out by the central government in case of financial distress. This moral hazard problem is known as the soft-budget constraint problem (Kornai et al. 2003). It has been studied, among others, by Velasco (2000), Rodden et al. (2003), and Krogstrup and Wyplosz (2010).

During election campaigns, local politicians may adopt opportunistic fiscal policies to signal competence and woo the electorate (Rogoff and Sibert 1988). Evidence on political business cycles at the subnational level is extensive. For developing countries, see Sakurai and Menezes-Filho (2011) for Brazil, Drazen and Eslava (2010) for Columbian municipalities, and Khemani (2004) for Indian states. For analyses of the rents appropriated while being in office see Battaglini and Coate (2008), and Yared (2010).

Government fragmentation and political instability may also be associated with larger deficits (Roubini and Sachs 1989) and debt can be used strategically (Persson and Svensson 1989; Alesina and Tabellini 1990) by a government that expects to lose the next election to a party with different ideology, in order to condition the latter's policy options. For evidence at the local level, see Pettersson-Lidbom (2001) for Sweden, and Solé-Ollé (2006) for Spain.

Conflicting interests among different generations regarding fiscal policy management were analyzed by Song et al. (2012). They developed a dynamic politico-economic model of government debt where debt is used by governments to shift the fiscal burden to future generations.

4.2 Balanced Budget Rules

Recognizing the importance of promoting sound fiscal policy at the subnational level, several countries have imposed numerical targets for subnational governments' budgets, namely balanced-budget rules. Pioneering studies to evaluate their efficacy in securing fiscal discipline where performed in the US (Holtz-Eakin 1988; Poterba 1994; Bohn and Inman 1996). Results indicate that balanced-budget rules improve fiscal outcomes, particularly on the part of the budget subject to constraints. However, fiscal rules are frequently hard to enforce, and they can generate incentives for governments to circumvent them through creative accounting practices (Milesi-Ferretti 2003; Beetsma et al. 2009), such as the substitution of spending on items subject to numerical rules with those that are not. Focusing on the European Union, Von Hagen (2006) finds that the fiscal rule of the Maastricht Treaty significantly diminished political budget cycles, but had small effects in restricting fiscal policy in larger countries. He also concludes that rules have been more effective in countries with good budgeting institutions. For Japan, Von Hagen (2006) concludes that rules had little effectiveness.

Keywords and Definitions

Ability to pay principle	A principle of taxation stating that taxes should be levied according to the taxpayers' ability to pay, those who have more wealth or earn more money should pay more taxes
Benefit principle	A principle of taxation stating that those who benefit the most from public goods/services should pay more to support those goods/services
Budget	Document or a collection of documents comprising a detailed description of the expected revenues and expenditures of a given institution, associated with the activities that are planned for achieving certain purposes or goals, within a given period
Participatory budgeting	A process of direct participation of citizens in deliberations and negotiations over public spending and resource allocation of a defined public budget
Performance budget	Budget that associates the input of resources with the service outcomes or results
Tax	A compulsory contribution to public revenue levied by the government on income, or property, or added to the cost of some goods, services and transactions
User charge	Sum of money demanded by a government for the use of a publicly provided good or service

References

Alam M (ed) (2010) Municipal infrastructure financing. Innovative practices from developing countries. Commonwealth Secretariat Local Government Reform Series No. 2. London
Alesina A, Tabellini G (1990) A positive theory of fiscal deficits and government debt. Rev Econ Stud 57(3):403–414
Battaglini M, Coate S (2008) A dynamic theory of public spending, taxation and debt. Am Econ Rev 98(1):201–236
Beetsma R, Giuliodori M, Wierts P (2009) Planning to cheat: EU fiscal policy in real time. Econ Policy 24(60):755–804
Bird RM (2012) Subnational taxation in large emerging countries: BRIC plus one. Institute on Municipal Finance and Governance Paper on Municipal Finance and Governance, 6. Munk School of Global Affairs. IMFG, Toronto
Bohn H, Inman RP (1996) Balanced-budget rules and public deficits: evidence from the U.S. states. Carnegie-Rochester Conference Ser Public Policy 45:13–76
Drazen A, Eslava M (2010) Electoral manipulation via voter-friendly spending: theory and evidence. J Dev Econ 92:39–52
Fisher R (2007) State and local public finance, 3rd edn. Thomson South-Western, Mason
Foster V (2005) Ten years of water service reform in Latin America: towards an Anglo-French model. Water Supply and Sanitation Sector Board Discussion Paper Series, 3. World Bank, Washington, DC
Guess GM, Leloup LT (2010) Comparative public budgeting. University of New York Press, Albany, NY
Hagemann R (2011) How can fiscal councils strengthen fiscal performance? OECD J: Econ Stud, 2011/1. http://dx.doi.org/10.1787/eco_studies-2011-5kg2d3gx4d5c. Accessed 13 Apr 2014
Hemming R, Joyce P (2013) The role of fiscal councils in promoting fiscal responsibility. In: Cangiano M, Curristine T, Lazare M (eds) Public financial management and its emerging architecture. IMF, Washington, DC, pp 205–224
Holtz-Eakin D (1988) The line item veto and public sector budgets: evidence from the states. J Public Econ 36:269–292
International Monetary Fund (2001) Government finance statistics manual. IMF, Washington, DC
Kaganova O (2011) Guidebook on capital investment planning for local government. World Bank Urban Development Series Knowledge Papers, 13. The World Bank, Washington, DC
Kelly R (2002) Designing a property tax reform strategy for sub-Saharan Africa: an analytical framework applied to Kenya. Public Budg Finance 20(4):36–51
Khemani S (2004) Political cycles in a developing economy. J Dev Econ 73:125–154
Kopits G (2011) Independent fiscal institutions: developing good practices. OECD J Budg 11 (3):35–52
Kornai J, Maskin E, Roland G (2003) Understanding soft budget constraint. J Econ Lit 41:1095–1136
Krogstrup S, Wyplosz C (2010) A common pool theory of supranational deficit ceilings. Eur Econ Rev 54(2):273–281
Kurian M (2010) Financing the millennium development goals for water and sanitation: issues and options. In: Kurian M, McCarney P (eds) Peri-urban water and sanitation services: policy, planning and method. Springer, Dordrecht
Kurian M, Ardakanian R (eds) (2015) Institutional arrangements and governance structure that advance the nexus approach to management of environmental resources. In: Governing the nexus: water, soil and waste resources under conditions of global change. Springer, New York
McCluskey WJ, Cornia GC, Walters LC (eds) (2012) A primer on property tax: administration and policy. Blackwell Publishing Ltd., Oxford

McCluskey WJ, Trinh H (2013) Property tax reform in Vietnam: options, direction and evaluation. Land Use Policy 30:276–285

Mikesell JL (2007) Fiscal administration in local government: an overview. In: Shah A (ed) Local budgeting, public sector governance and accountability series. The World Bank, Washington, DC, pp 15–51

Milesi-Ferretti G (2003) Good, bad or ugly? On the effects of fiscal rules with creative accounting. J Public Econ 88:377–394

Mullins D (2007) Local budget process. In: Shah A (ed) Local budgeting, public sector governance and accountability series. The World Bank, Washington, DC, pp 213–264

Nauges C, van den Berg C (2008) Economies of density, scale and scope in the water supply and sewerage sector: a study of four developing and transition economies. J Regul Econ 34 (2):144–163

Organisation for Economic Co-operation and Development (2012) Principles for public governance of public private partnerships. OECD, Paris

Persson T, Svensson L (1989) Why a stubborn conservative would run a deficit: policy with time-inconsistent preferences. Quart J Econ 104(2):325–345

Pettersson-Lidbom P (2001) An empirical investigation of the strategic use of debt. J Polit Econ 109(3):570–583

Poterba J (1994) State responses to fiscal crises: the effects of budgetary institutions and politics. J Polit Econ 102(4):799–821

Reingewertz Y (2012) Do municipal amalgamations work? evidence from municipalities in Israel. J Urban Econ 72:240–251

Reschovsky A (2007) Compensating local governments for differences in expenditure needs in a horizontal fiscal equalization program. In: Boadway R, Shah A (eds) Intergovernmental fiscal transfers. Principles and practice. The World Bank, Washington, DC, pp. 397–424

Rodden E, Eskeland GS, Litvack J (eds) (2003) Fiscal decentralization and the challenge of hard budget constraint. The MIT Press, Cambridge

Rogoff K, Sibert A (1988) Elections and macroeconomic policy cycles. Rev Econ Stud 55:1–16

Roubini N, Sachs J (1989) Political and economic determinants of budget deficits in the industrial democracies. Eur Econ Rev 33:903–938

Sakurai SN, Menezes-Filho N (2011) Opportunistic and partisan election cycles in Brazil: new evidence at the municipal level. Public Choice 148(1–2):233–247

Schroeder L (2007) Forecasting local revenues and expenditures. In: Shah A (ed) Local budgeting. Public sector governance and accountability series. The World Bank, Washington, DC, pp. 53–77

Sepulveda C, Martinez-Vazquez J (2011) Explaining property tax collection in developing countries: the case of Latin America. International Studies Program Working Paper No. 11-09. Andrew Young School of Policy Studies, Georgia State University, Atlanta

Shah A, Shah S (2006) The new vision of local governance and the evolving roles of local governments. In: Shah A (ed) Local governance in developing countries. Public sector governance and accountability series. The World Bank, Washington, DC, pp. 1–44

Shah A (2007a) Local budgeting. Public sector governance and accountability series. The World Bank, Washington, DC

Shah A (2007b) Participatory budgeting. Public sector governance and accountability series. The World Bank, Washington, DC

Solé-Ollé A (2006) The effects of party competition on budget outcomes: empirical evidence from local governments in Spain. Public Choice 126:145–176

Song Z, Stosletten K, Zilibotti F (2012) Rotten parents and disciplined children: a politico-economic theory of public expenditure and debt. Econometrica 80(6):2785–2803

Velasco A (2000) Debt and deficits with fragmented fiscal policy-making. J Public Econ 76:105–125

Von Hagen J (2006) Fiscal rules and fiscal performance in the European Union and Japan. Monetary and Economic Studies, Institute for Monetary and Economic Studies, Bank of Japan 24(1):25–60

World Bank (2004) World development report: making services work for the poor. World Bank, Washington, DC
World Bank (2013) Beyond the annual budget: global experience with medium-term expenditure frameworks. World Bank, Washington, DC
Yared P (2010) Politicians, taxes and debt. Rev Econ Stud 77:806–840

Chapter 3
Issues of Accountability and Policy Outcomes: Key Lessons from Recent International Experience

Abstract This section focuses on the relationship between accountability and policy outcomes such as the quality of service delivery. This relationship, the main dimensions of accountability and several important contextual factors are analyzed in the context of decentralization. Then, this section focuses on the recent experience in alternative service delivery and financing models. The role of indicators/indices for policy, program, and project design and for monitoring results is also discussed. Finally, the last section deals with the lessons from the international experience and implications for the nexus approach to management of environmental resources.

Keywords Accountability · Governance · Public-private partnership · Results-based financing

This section focuses on the relationship between accountability and policy outcomes. An introductory general discussion of the above-referred relationship is followed by a description of the main dimensions of accountability and of how they can be enhanced at the subnational government level. Then, the analysis focuses on the contextual factors influencing the design and outcomes of decentralization and on the effects of the latter on the quality of governance. Taking into account that decentralization is not the only available alternative to centralized public service provision, the following section focuses on the recent experience in service delivery and financing models, including public–private partnership arrangements (e.g., management and operating, leases, concessions, build-own-transfer, divestiture), independent providers, community-managed networks, etc., and financing models, such as results-based financing, output-based aid, and budget support. The role of indicators/indices for policy, program, and project design and for monitoring results is also discussed. Finally, the last section deals with the lessons from the international experience and implications for the nexus approach to management of environmental resources.

L.G. Veiga et al., *Intergovernmental Fiscal Relations*,
SpringerBriefs in Environmental Science, DOI 10.1007/978-3-319-06296-9_3

1 Accountability, Governance, and Public Expenditure Efficiency

As thoroughly explained and illustrated in the 2004 *World Development Report*, the quality of service delivery varies considerably across developing countries, and there is a general tendency for services to "fail poor people—in access, in quantity, in quality" (Word Bank 2004, p. 1).[1] Strengthening government accountability and amplifying poor people's voice could contribute to better governance and, consequently, to a more efficient and egalitarian use of public resources.[2] Based on the experience with service delivery, the *World Development Report* suggests a constellation of ways to make services work for the poor, taking into account the service and country or region characteristics. The adequacy of each of the eight possibilities (sizes) presented in Fig. 1 depends on how much the political system of the country is geared toward pro-poor public services and on how homogeneous the clients are.

Decentralization is recommended when policies are pro-poor and clients are heterogeneous (cases 3 and 4). If services are subject to capture, and policies are not pro-poor (cases 5 to 8), then client power should be reinforced as much as possible. When political circumstances, or a variety of other factors, make governance reform very difficult at the national level, there may still be room for improvement by taking advantage of local level entry points (World Bank 2007b). That is, a country's overall quality of governance and service delivery could still be improved by providing greater discretion and accountability to subnational governments (World Bank 2007a). With increased accountability and more pro-poor policies, situations formerly classified in cases 7 and 8 of Fig. 1 could become more in line with cases 3 and 4, thus justifying greater involvement of subnational governments.

Theory suggests that decentralization can improve public expenditure efficiency and generate better policy outcomes,[3] since the greater proximity of subnational governments gives citizens more influence over local officials, promotes competition among subnational governments, increases accountability, and reduces corruption. In practice, however, decentralization may actually worsen outcomes when

---

[1] Focusing on water supply and sanitation (WS&S) services, Allen et al. (2006, p. 13) claim that "it is a well-documented fact that in many countries water supply and sanitation systems fall short of present and future requirements. Some of the world's poorest people lack adequate access to these most basic of services." In fact, one of the United Nations Millennium Development Goals is to halve the proportion of population without access to clean water and sanitation by 2015. Fulfilling this goal will require considerable investments by developing countries and donor agencies.

[2] According to the Third World Water Forum Statement (http://www.worldwatercouncil.org/), "Many countries face a governance crisis rather than a water crisis. Good water governance requires effective and accountable sociopolitical and administrative systems adopting an integrated water resources approach with transparent and participatory processes that address ecological and human needs."

[3] Recall Volume I. Trends in financing of public services.

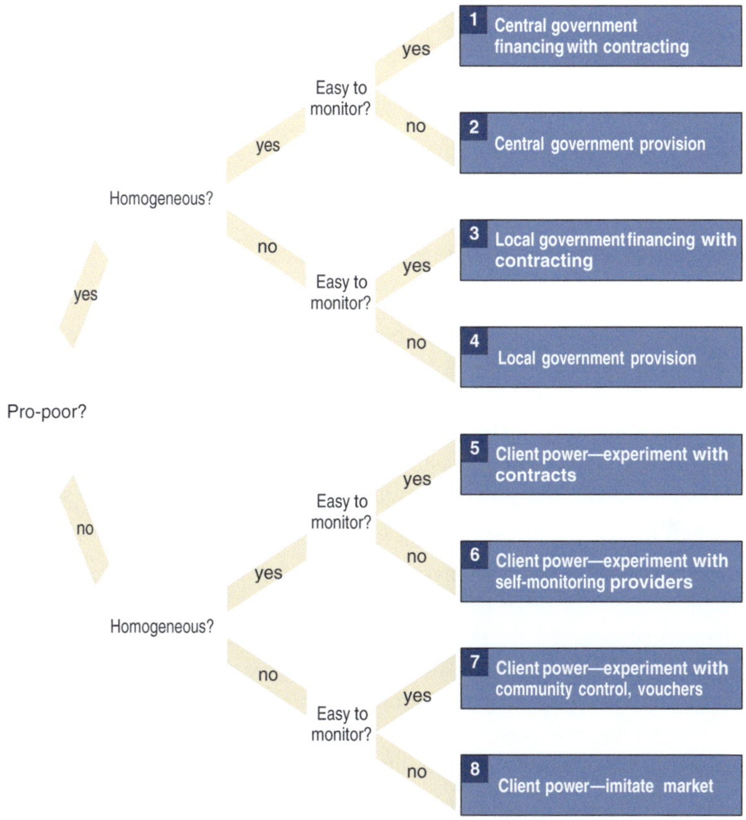

Fig. 1 Eight sizes fit all? *Source* Word Bank (2004, p. 14)

subnational governments do not have the means, capacity, or incentives to act as theory predicts. That is, the outcomes of decentralization depend heavily on how the latter is designed and on the local context.

The following sections focus on how local accountability can be improved, on the governance effects of decentralization in several countries, and on what the recent international experience teaches us regarding the relationship between accountability, governance, and policy outcomes.

2 Decentralization and Local Accountability

Decentralization involves giving subnational governments a reasonable degree of discretionary powers, so that they are able to respond to citizens' needs. However, in the absence of an appropriate degree of accountability, greater discretion may actually lead to the misuse and abuse of the new powers, to the capture of power by local elite groups, and to poor service delivery. Thus, decentralization should be accompanied by a reinforcement of subnational governments' accountability in a way that creates incentives for better performance. This section will focus on the different dimensions of accountability in local context and on the approaches to reinforce it.[4]

2.1 Dimensions of Accountability

Accountability involves a relationship between a principal and an agent. Regarding service delivery, the actors involved are the citizens-clients, the policymakers, and the providers (organizational and frontline). Decentralization adds subnational governments to this list of actors (see Fig. 2).

Accountability is then a relationship in which one or more of these actors is/are able to keep in check the power of another actor (or actors). Four sets of relationships apply:

- Vertical accountability: of policymakers and state organizations to citizen-voters through elections or other political channels;
- Horizontal accountability: of one state institution to another at the same level (e.g., control of the legislative branch over the executive);
- Upward accountability: of a subnational government to the national government or its overseeing agencies;
- Downward accountability: of the subnational government to its citizens.

Regarding what the agent (subnational government) is accountable for, three categories apply:

- Political accountability: to the citizen-voters on the way policy is conducted and services are delivered;
- Administrative accountability: regards the administrative decisions of local officials on issues such as recruitment, procurement, and land use and planning;
- Fiscal/financial accountability: regards the management of local finances and its outcomes.

The object of accountability can also take into account the process (compliance with rules) or the outcome (quality of service delivery) of decisions or actions taken by subnational governments. Regardless of the object of accountability, there are essentially two approaches to promote it:

- Public or supply-side accountability: refers to the institutionalized practice of account giving, the obligation of public authorities to explain how they are carrying out their responsibilities;

[4] This section's exposition is essentially based on Word Bank (2007a) and Yilmaz et al. (2010).

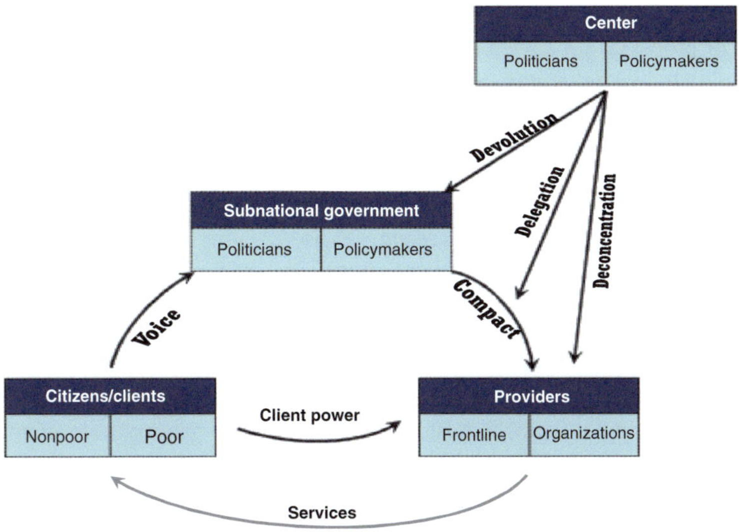

Fig. 2 Actors of accountability. *Source* World Bank (2007a, p. 7)

- Social or demand-side accountability: based on citizen oversight, in the sense that citizens or civil society organizations demand accountability.

The subsection below explains how local political, administrative, and fiscal discretion can be given and accountability strengthened using the two approaches referred to above.

2.2 *Strengthening Political, Administrative, and Financial Accountability*

Downward political accountability can more easily be attained in the presence of free and fair elections and of local political competition (with multiple parties and candidates) for mayors and councilors. Clear separation of powers between the executive and legislative branches, and independent courts are also needed. Political accountability can be reinforced by improving the quality of the electoral system allowing independent candidates to run in local elections, limiting time in office, reserving seats in local councils to minority/vulnerable groups, allowing recall of elected officials, increasing the transparency of election and campaign financing, and by strengthening the oversight role of the legislative branch (council) over the executive. Besides these public or supply-side measures, political accountability can also be reinforced with demand-side interventions. These would involve approving legislation that allows citizens to demand explanations and justifications from subnational governments (e.g., public hearings, public petitions, administrative complaints, and the right to initiate a recall or a referendum) and the institutionalization of formal bodies or processes for citizen oversight.

Administrative autonomy of subnational governments involves some degree of discretion in the areas of legislation and regulation, local procurement, and local employment. Then, administrative accountability implies that local civil servants are accountable to their top administrative officers and to outside bodies on the way they exercise their discretion. Public or supply-side measures to enhance administrative accountability involve, among others, creating/reinforcing independent judicial/quasi-judicial agencies that investigate cases of misconduct and corruption; external audits by independent agencies; administrative courts, which evaluate the compliance of decisions and actions with rules; procurement rules, processes, and independent audits; standards for service delivery; rules regarding civil service practices; flexible and performance-oriented career management; and service quality monitoring systems/indicators. A demand-side approach to reinforcing administrative accountability involves greater citizen participation in decision-making processes (regarding projects, procurement, hiring, etc.) and the creation of public oversight mechanisms, which enable citizens to monitor local service delivery and to express their opinion on its quality.

Financial accountability implies transparency and prudency in the management of public funds and that subnational governments report on what they have achieved with their expenditures. This requires local fiscal autonomy, clear rules for intergovernmental transfers (which should be public information), transparent local public audit systems, and clear rules regarding local hard budget constraints and borrowing. Besides these public or supply-side measures to improve financial accountability, the latter can be attained by social or demand-side measures, such as making publicly available detailed subnational government financial information; introducing/reinforcing participatory budgeting practices, independent budget analyses, and public expenditure tracking programs; and through gender-sensitive planning, budgeting and resource allocation.

3 Decentralization and Governance

Good subnational governance is associated with high degrees of decision-making autonomy (discretion) and accountability. Unfortunately, most developing countries start(ed) off with very low discretion and accountability, the worst-case scenario. Gradually increasing both discretion and accountability would be the desirable strategy, but that seldom happens in the real world. Countries such as Indonesia adopted a trajectory that focused on discretion (through big bang decentralization) in the short run and strengthened the accountability of subnational governments later on. However, new administrative and fiscal powers are unlikely to be well used if not disciplined by administrative and political mechanisms. Thus, the major risk associated with this strategy is that subnational governments misuse and abuse of the new powers or these are captured by local elites before accountability is sufficiently increased.

Ex-communist transition economies in Eastern Europe adopted a strategy that placed greater priority on promoting political, administrative, and financial accountability in the short run, and increased the autonomy of subnational governments later on. The major problem with this strategy is that, in the short run, subnational governments may lack the financial means and decision-making autonomy necessary to provide good quality service delivery to their citizens. The latter may become disillusioned with decentralization and disengage from local democracy if they feel they receive insufficient benefits. This trajectory's main advantage is that the risk of elite capture and corruption is smaller than under a trajectory that increases discretion first and accountability later on.

Diverse contextual factors can heavily influence the trajectory chosen, the shape decentralization takes and the outcomes it delivers in a specific country. These include, among others, demographic and social characteristics, political economy factors, size of the economy, the degree of democratization, and aid dependence (LDI 2013). Besides influencing service delivery outcomes, decentralization can also affect governance in several dimensions, such as the size of government, democratization, corruption, and political instability. These matters will be analyzed below.

3.1 Contextual Factors Influencing the Design and Outcomes of Decentralization

Decentralized service provision is expected to enhance the quality, efficiency, and equity of service delivery, given the greater proximity of subnational governments to citizens. However, in practice, unless decentralization brings along greater accountability and governance, citizens may not obtain better services. The paragraphs below briefly review studies that argue/find that low levels of accountability and governance related to, or resulting from, the social, political, or economic context, institutional design, political economy issues, and local capacity negatively affect the quality of services delivered by subnational governments.

Regarding the social/political/economic context, the empirical literature tends to find strong support that context matters for the success or failure of decentralization. Some qualitative studies argue that a participatory political culture is necessary for the efficacy of the mechanisms adopted for citizen participation in decision-making.[5] In the field of water supply and sanitation (WS&S), the WELL's (1998) guidance manual recommends the participation of primary and secondary stakeholders and that their perspectives and knowledge are incorporated in any development.

[5] Golooba-Mutebi (2005) study health decentralization in Uganda. Regarding citizen participation in this country, see also World Bank (2007a).

Furthermore, they consider that the full involvement of users in decision-making is of special importance in WS&S services. Kurian and Ardakanian (2013) also stress the importance of participation for the advancement of the nexus approach to the management of water, waste, and soil resources.[6] They argue that it is necessary to engage with consumers to ascertain their views on the reliability, affordability, and adequacy of services.

Although there are studies that go further by suggesting that decentralization requires a democratic framework to deliver the desired results,[7] service delivery can also improve in less democratic regimes, as it did under Suharto in Indonesia. A country's level of economic and institutional development also has an important role, as Jütting et al. (2004) find that countries where the central government is not able to perform its basic functions, decentralization worsens service delivery to the poor.

Several studies indicate that institutional design is a crucial factor affecting decentralization performance. Some focus on the types and extent of functions being devolved to subnational governments, indicating problems of institutional incoherence related to partial devolution of authority, overlapping mandates of local stakeholders, unclear responsibilities of subnational governments, and lack of coordination. In this respect, Allen et al. (2006) report that institutional fragmentation tends to be severe in the metropolitan context, and that weak links and limited municipal power in sectors such as transport, water, energy, solid and liquid waste management, and land use planning often result in uncertainty as to which institution administers which specific area or activity. Kurian and Ardakanian (2013) argue that environmental governance in developing and emerging economies suffers from fragmented approaches to planning and policy implementation, which are due to competition among urban and rural subnational governments for central government transfers, overlapping jurisdictional boundaries, and lack of coordination.

Other studies point to the importance of local finances, arguing that problems of insufficient assignment of financial resources to subnational governments or low capacity of the latter to generate own revenues, have originated negative effects of decentralization on service delivery. For example, Dinar et al. (2007) find that the success of decentralization in water resource management (in 83 river basins worldwide) is positively associated with the local share and discretion over central government funding and with the share of users paying tariffs.

A number of studies look at the institutional mechanisms for local transparency and accountability (mainly, to citizens), and find that they are essential ingredients of successful decentralization.[8] In fact, when these mechanisms are ineffective,

[6] This approach recognizes that "environmental quality can be satisfied only if soil, waste and water resources are managed in a sustainable and integrated manner" (Kurian and Ardakanian 2013, p. 56).

[7] See Faguet (2014) for a survey on the relationship between decentralization and governance.

[8] De Grauwe et al. (2005) find that weak accountability is a major obstacle to improving decentralized education in sub-Saharan Africa. World Bank (2007a) analyzes local government discretion and accountability in Uganda, Kerala (India), and in the Philippines.

corrupt practices by local officials lead to poor service delivery. Baltaci and Yilmaz (2006) point out several shortcomings with regard to financial accountability at the subnational government level in several countries. They argue that, in order to reduce the resulting misuse of public resources, fraud, and corruption, it is necessary to reinforce both financial and social accountability mechanisms. One way to do it is to rely more frequently on participatory budgeting (PB). Gonçalves (2014) finds that Brazilian municipalities adopting PB increased spending in health and sanitation and reduced infant mortality significantly more than the municipalities that did not adopt PB. Davis (2004) presents empirical information regarding the types and magnitude of corrupt behaviors documented in water supply and sanitation (WS&S) service provision in South Asia, where its quality is generally low.

Political economy issues also affect decentralization performance, as emphasized by the literature on second-generation fiscal federalism, which takes into account the fiscal and political incentives facing subnational officials (Weingast 2014). A number of studies argue that central–local relationships help explain the malfunctioning of decentralization in several countries, as several governments use decentralization as a tool for consolidating party interests rather than improving service delivery. Agrawal and Ribot (1999) argue that central governments in South Asia and West Africa are not genuinely interested in more local autonomy in forest management. For example, the central government of Uganda is said to use creation of new districts as a form of patronage (Word Bank 2007a). Weingast (2014) points out that predatory central governments are a problem throughout the developing world, and they hinder the operation of an otherwise well-designed federal system. Local dynamics are also very important, as service quality suffers when local elites are able to capture the new powers assigned to subnational governments (Bardhan and Mookherjee 2000).

Finally, another set of studies identifies local capacity issues as an important determinant of decentralization performance, in the sense that subnational governments should be able to assume the new responsibilities in a competent way,[9] and local citizens must be able to engage with subnational governments, hold them accountable, and participate in local decision-making.[10] In fact, successes of the decentralized approach in water supply, sanitation, and irrigation, in countries such as Mexico and Egypt, are related to increases in user involvement in management.[11] Local participation, capacity building, and stakeholder forums are key ingredients of successful decentralization of natural resource management (see Litvack and Seddon 1999). According to WELL's (1998) guidance manual, opportunities for

[9] Gómez (2008) argues that health was decentralized too quickly in Brazil, which undermined local ability to manage services. According to World Bank (2007a), the sub-country level in Uganda is chronically under-resourced and the district governments' taxing potential is not commensurate with the responsibilities being devolved.

[10] Jones et al. (2007) find that illiteracy and lack of information, especially among the most vulnerable groups, hinder inclusive participation in India.

[11] Successful water supply and sanitation programs are also under implementation in countries such as Benin, Bolivia, Ecuador, Ghana, India, and South Africa (see Litvack and Seddon 1999).

institutional strengthening in WS&S projects include capacity building in the subnational government organizations and in NGOs, and the development of stronger links between the two. It also involves empowerment of primary stakeholders and the development of people's capabilities at local level. Finally, according to Kurian and Ardakanian (2013), the limitations of conventional technological transfer in terms of addressing environmental challenges led to the acknowledgement that capacity development approaches that aim to facilitate technology adaptation are a necessary ingredient of the sustainable and integrated management of water, soil, and waste resources. In fact, the implementation of the nexus approach is complex and requires the enhancement of capabilities and knowledge of developing countries public officials.

3.2 The Impact of Decentralization on Governance

According to Faguet (2014, p. 2), "the most important theoretical argument concerning decentralization is that it can make government more accountable and responsive to the governed." This section summarizes the main results regarding issues such as expenditure composition, size of government, democratization, corruption, and reduction of political instability.

3.2.1 Size of Government and Expenditure Composition

Brennan and Buchanan (1980) view the public sector as a "Leviathan" that seeks the maximization of tax revenues, whose actions could be constrained by fiscal decentralization, given the increase in tax competition in the context of inter-jurisdictional mobility of taxpayers. Thus, the size of the public sector should vary inversely with the degree of decentralization. A dissenting view is presented by Oates (1985), who argues that losses in economies of scale, poorer quality of local bureaucrats (relative to national ones), and soft budget constraints of subnational governments, may imply that decentralization generates greater sizes of government.

The results of empirical studies are not conclusive, as no consistent evidence has been found to support or reject the Leviathan hypothesis. While Zax (1989) finds supporting evidence, Oates (1985) and Nelson (1987) reject the Leviathan hypothesis. When the size of government is evaluated in terms of public sector employment, there is a tendency to find that decentralization increases it. In a cross-country study, Martinez-Vazquez and Yao (2009) find that the increase in public employment at the subnational government level more than compensates for the decrease at the central level.

Regarding the composition of public expenditures, studies using panel data covering several countries, such as Arze del Granado et al. (2005), find strong evidence that fiscal decentralization increases the share of education and health expenditures in total government expenditures, especially in developing countries.

3.2.2 Democratization

There is some evidence of a mutually reinforcing relationship between decentralization and democratic governance. On the one hand, political decentralization, through local elections and participation, can improve local accountability and, consequently, governance (Blair 2000). On the other hand, democratization in many countries of Africa, Eastern Europe, and Latin America provided considerable impetus to decentralization.

However, the chain of causality from decentralization to governance must involve political competition, as it motivates politicians to offer better public services at lower corruption-price (Myerson 2014). Furthermore, according to Weingast (2014), the form of the party system is the key to the positive effects of decentralization on governance. Ideally, the party system should balance the power of national and local elites, forcing them to cooperate in order to win elections at both levels. If, on the other hand, one level clearly dominates the other, decentralization will not produce the desired effects.[12]

Even in some cases where party competition is insufficient, greater civic engagement and participation in local decision-making could still lead to better governance. Díaz-Cayeros et al. (2014) examine the traditional collective decision-making process known as *usos y costumbres* in Oaxaca, Mexico. They find that poor indigenous communities that use this form of self-management and participatory democracy enjoy higher levels of civic engagement and governance than similarly poor communities do where political parties compete for power. Thus, what matters most may not be the existence of political parties, but if subnational governments are effectively democratically accountable to citizens. In this sense, decentralization requires democracy in order to achieve its potential (Faguet 2014).

3.2.3 Corruption

Decentralization has the potential to reduce corruption by enhancing accountability and competition among subnational governments, by providing exit and voice mechanisms and higher levels of information and transparency, or because of lower expected gains from corruption with higher probability of detection and punishment (Boadway and Shah 2009). However, decentralization may have the opposite effect due to weaker monitoring, controls and audits, involvement of a larger number of officials, higher incidence of clientelism and elite capture at the local level, and to higher incentives related to lower salaries and poorer career prospects (Bardhan and Mookherjee 2000; Shleifer and Vishny 1993).

Despite the theoretical arguments pointing in both directions, most of the recent empirical literature finds that decentralization lowers corruption (e.g., Fisman and

---

[12] See Ardanáz et al. (2014) for a study on Argentina, whose federal system is characterized by strong political "baronies" in some poorer and less-developed provinces.

Gatti 2002; Ivanyna and Shah 2011, 2013). Nevertheless, this is not a settled issue, as there are some empirical studies concluding that decentralization increases corruption (Fan et al. 2009; Treisman 2000).

3.2.4 Political Instability

Decentralization may relieve political tensions and reduce risks of violence and secession in heterogeneous countries with strong subnational (ethnic, regional, or religious) identities. The devolution of power and resources to subnational governments allows the latter to implement policies better suited to their citizens' specific interests, thus reducing their grievances and level of discontent. Additionally, the transfer of power to local leaders willing to settle for increased local autonomy reduces the support for more radical local politicians willing to promote violence and secession (Diamond et al. 1995; Hechter 2000).

Although it is also possible that federalism exacerbates the risk of regional secession, by institutionalizing social or ethnic cleavages and establishing the formation of regional parties that might drive a secessionist agenda, that risk can be reduced by creating smaller provinces (or other local/regional units) and by choosing parliamentary over presidential democracy at both national and subnational levels.[13]

4 Recent Experience in Service Delivery and Financing Models

As suggested by World Bank (2004), it is necessary to strengthen accountability in the key relationships among the actors involved in service delivery (illustrated in Fig. 2). This can be done by separating the policymakers from the providers, and making the latter more responsive to clients. Dispersing ownership through decentralization and private participation, promoting competition through benchmarking, using independent providers to guarantee alternative access, and charging for services are ways of strengthening compacts, client power and voice. Linking public funding and official development aid (ODA) to outcomes/results through results-based financing (RBF) may also contribute to better quality of service provision.

[13] Myerson (2014) suggests that no province should comprise more than 20 % of the national population. He also argues that parliamentary systems are less likely to suffer democratic breakdown than presidential systems, especially in more heterogeneous countries.

4.1 Recent Trends in Service Delivery Models

Decentralization, or devolution to local governments, is by no means the only available alternative to central government provision. Other commonly used options include contracting out to the private sector and NGOs, selling concessions to the private sector, community participation, and direct transfers of resources and responsibilities to households.

Private sector participation has grown significantly since the 1990s in many forms and across many regions, with the hope that public-private partnership (PPP) arrangements would turn around poorly performing public provision by bringing new expertise, financial resources, and a more commercial orientation. In several developing countries where fiscal pressures were intense, the prospect of shifting investment responsibility to private infrastructure providers also played an important role in the increased acceptance of private sector involvement. Private participation is also a way of separating policymakers and service providers through compacts and voice (World Bank 2004). First, when designing compacts, private providers usually require explicit contracts defining their responsibilities and those of policymakers. The voice channel can also be strengthened if all stakeholders are involved in the process. In fact, service delivery standards and services for poor people are often explicit in the policy debate on private participation in service delivery.

Figure 3 illustrates the extent of private sector participation across a variety of PPP arrangements. These stand between the extreme cases of decentralization, where the public sector owns and operates assets, to full divestiture (privatization), where the assets are sold to a private entity, which provides services directly to consumers. Common forms of PPPs, implying increased private sector participation, include management and operating (M&O) contracts, leases/affermages, concessions, Build-Own-Transfer (BOT), Build-Own-Operate (BOO), or Design-Build-Operate (DBO) projects, and joint ventures.[14]

The total amount of private investment in infrastructure, its sectorial composition, and the total number of new PPP projects in middle and low income countries are shown in Fig. 4. Most of the investment has been concentrated in energy and telecom, transport has gained some weight after 2005, while the water and sewage sector always had a residual share. While Greenfield projects (usually involving upfront capital investments in the form of BOTs, BOOs, DBOs, etc.) and divestitures (privatizations) are responsible for most of the private participation in the energy, telecom and transport sectors, concessions, leases, and M&O contracts are

[14] Other possibilities include, among others, Cooperative and twinning contracts, Build-Lease-Transfer (BLT), Build-Own-Operate-Transfer (BOOT), Design-Build-Finance-Operate (DBFO), Design-Build-Operate-Transfer (DBFT), Design-Construct-Manage-Finance (DCMF). For more detailed information on PPPs, see the World Bank's PPP Resource Center Website (www.worldbank.org/ppp).

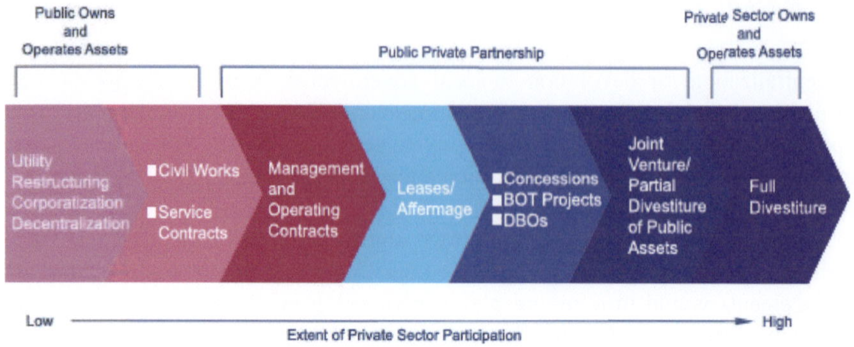

Fig. 3 Extent of private sector participation. *Source* www.worldbank.org/ppp

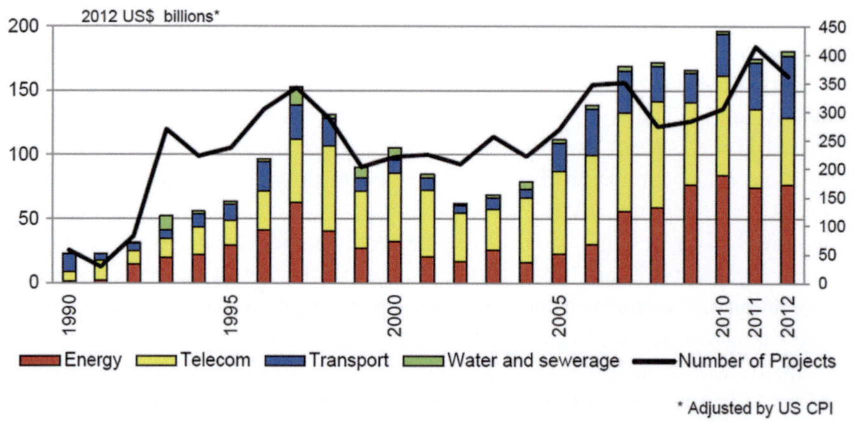

Fig. 4 PPPs in middle and low income countries. *Source* Word bank and PPIAF, PPI project database.

more common models in the water and sewage sector, where public ownership and not divestiture are the norm.

Compared to the other sectors, private participation in infrastructure (PPI) in the water and sewage sector has been limited and disappointing in relation to the high expectations prevailing in the 1990s. In fact, it seems to be an unpredictable source of finance, given the number of problems encountered with even the relatively limited number of transactions completed.[15] Experience suggests that PPI has more potential to improve efficiency and quality of service than to mobilize new finance (Annez 2006; Marin 2009). The much lower private participation in water and sewage projects

[15] Marin (2009) argues that PPP projects in the water sector have been controversial, especially after a series of highly publicized contract terminations raised serious doubts about their suitability for developing countries.

may be related to archaic land tenure arrangements or inadequate guarantees that people will pay for services because of political interference (Kurian 2010).[16] In fact, solid legal and policy frameworks are essential to the attraction of private investment, as they play an important role in guaranteeing that costs will be recovered, by defining service and environmental standards and providing guidelines for tariff structures, taxes, and government or ODA transfers. Thus, as argued in OECD (2009), attracting sustainable flows of finance depends on thorough reforms in this sector's governance and on using strategic financial planning with a long-term perspective.

Relying more on small independent providers may also improve urban water services (World Bank 2004). Their organization varies from household vendors of water, small network providers, and private entrepreneurs to cooperatives. In some cases, they are primary suppliers, while in others they supplement the formal provider. Enhancing their role strengthens client power and, consequently, accountability in service provision. This can be done by granting them legal status, by not giving monopoly power to network providers, by enabling partnership and contracting between independent and network providers, and by allowing poor people more access to independent providers.

Community-managed networks/systems became more common in rural areas, as several governments realized the failures of top-down institutions.[17] Communities have been involved in the design and management of their water systems, paying for operations and maintenance costs. Although this client-based model has the virtue of placing the client at the center of the accountability relationship, it faces several challenges: technical support is required in the medium to long run; it faces difficulties in replenishing capital investments or charging higher tariffs to cover rising costs; problems of elite capture may occur in heterogeneous communities; and efficient technologies, which require scale economies, are not selected because of the common focus on village-level associations. Some of these problems could be mitigated with the financial and institutional support of local governments. Communities can also contract with a third party to manage local network systems.

4.2 Results-Based Financing

Results-based financing (RBF) is a funding model in which payments are only made after previously agreed upon results/outcomes can be proven to have been obtained. That is, funds are delivered after the funding entity has verified that some

[16] Working with local, rather than national governments, brings additional risks to private suppliers, related to political alliances and election calendars, and to the limited scope of authority of local governments (Annez 2006).

[17] Scholars such as Ostrom (1990) have argued that community-based organizations can lower transaction costs of monitoring access of common pool resources such as irrigation systems or forests. For a recent study of collective action in the Indian Himalayas, see Kurian and Dietz (2013).

agreed objectives have taken place. This, in turn, requires the funding to be focussed on objectives that can be clearly measured (and attained), in contrast to input-based approaches to paying for improved capacity, whether in the form of new technology, training, or equipment. RBF emerged as a key part of the "aid effectiveness" agenda in development financing before spreading to climate finance. Its proponents assert that it helps to allocate money more efficiently because the financial risks are transferred from donors to recipients, creating a much greater pressure on the latter to deliver on their promises. It is also expected to increase transparency, accountability, efficiency, private sector engagement and increased sustainability of public finances. The most common RBF approaches are output-based aid, conditional cash transfers, cash on delivery, and performance-based contracting.

Output-based aid (OBA) is a RBF mechanism that has been increasingly used to deliver basic infrastructure and social services to the poor.[18] It ties the disbursement of public funding to the achievement of clearly specified results in terms of improved access to basic services, such as water supply and sanitation, energy, health care, education, communications, and transportation. Outputs are defined as closely as contractually possible to the desired outcomes, and public funding is provided in the form of subsidies, which fill the gap between the total cost of providing a service to a user and the user fees charged. These are normally applied in three ways: one-off subsidies such as connection subsidies (which are the most common); transitional tariff subsidies that taper off as user contributions increase; and ongoing subsidies, which may be required where a continuous gap exists between affordability and cost recovery. The volume of World Bank Group OBA subsidy by region and sector is shown in Fig. 5. OBA has been mostly used in Latin America and Africa, and in the transport and health sectors. Its use in the water and sanitation sector is still relatively small (5 % of total), and is mostly concentrated in Africa.

Despite the donors' assertion that OBA, and RBF in general, improve aid effectiveness (see IDA 2009), some criticisms have been raised. It may distort development priorities, since many projects that contribute to social change are not easily quantifiable. The delivery of funding only after project implementation, and the high costs of data collection and auditing, may also limit the applicability of results-based approaches. Finally, according to Pereira and Villota (2012), results-based approaches are not particularly good at supporting aid effectiveness principles.

[18] There is an increased use of OBA by the World Bank, through the IBRD, the IDA (see IDA 2009) and the Global Partnership on Output-Based Aid (GPOBA). Other major results-based initiatives are the European Commission's Millennium Development Goals Contract (MDG-C), the GAVI Alliance (GAVI) Health System Strengthening support and Immunization Services Support; the Millennium Challenge Corporation's (MCC) Threshold and Country Programs, and the Global Fund to Fight AIDS, Tuberculosis and Malaria (GFATM). Pereira and Villota (2012) evaluate their effectiveness.

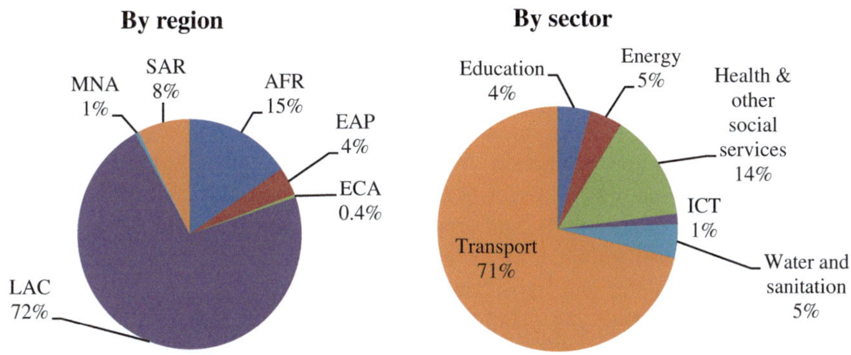

Fig. 5 Volume of OBA subsidy by sector and region in the WBG. *Source* http://www.oba-data.org/. Data obtained on February 18, 2014. *Note* Total subsidy = US$ 3,987 m. AFR = Sub-Saharan Africa; EAP = East Asia and Pacific; ECA = Europe and Central Asia; LAC = Latin America and Caribbean; MNA = Middle East and North Africa; SAR = South Asia region

4.3 Budget Support and Sector-Wide Approach

Budget support operations channel funds to national budgets and engage with fundamental policy reforms. They concern recurrent, predictable development policy lending (DPL) in a medium-term programmatic setting (see Iyer et al. 2005). In the scope of IDA assistance, budget support is designed to assist the design of a poverty reduction strategy credit (PRSC). Essentially, it offers the opportunity to scale up national programs toward the fulfillment of the Millennium Development Goals.

Budget support can also enable broad-based reform, improve the predictability of funding, remove externally imposed bottlenecks, reduce costs from fragmentation and separate project implementation units, and optimize the impact of both government and external finance.[19] Especially in countries that have made substantial progress in sector reform, it makes sense to move forward from sector-wide approaches (SWAPs) to budget support, as the latter allows for a broader development perspective prioritizing investment requirements and sector and regional priorities (see Kurian 2010).

[19] Focusing on the rural water supply, sanitation and hygiene (RWSSH) sector, Iyer et al. (2005) argue that when it is included as a focus or maintenance sector in budget support operations, the potential for significant progress is immense. But, when this is not the case, the sector risks being marginalized in the development policy debate and starved of investment funds.

4.4 The Role of Accountability, Governance, and Outcomes Indicators

From a development perspective, accountability and governance indicators can be used for monitoring and evaluation of governance programs and projects. In fact, output indicators are essential for RBF. They can also be used to establish benchmarks, objectives, targets, and goals in the development context. Regardless of the purpose for which indicators are used, three rules should be observed (UNDP 2007): (1) use a range of indicators instead of a single one; (2) use an indicator as a first question—not a last; and (3) understand an indicator before you use it.

Given the large number of indicators available, it is not possible to describe most of them in the present text. Therefore, only two of the most widely used and comprehensive databases of indicators will be referred to here[20]:

- *World Governance Indicators—WGI* (World Bank).[21] Provides indicators for six dimensions of governance: voice and accountability, political stability and absence of violence/terrorism, government effectiveness, regulatory quality, rule of law, and control of corruption.
- *Country Policy and Institutional Assessment*—CPIA (World Bank—IDA).[22] Rates 81 International Development Agency (IDA) recipient countries against a set of 16 criteria grouped in four clusters: economic management; structural policies; policies for social inclusion and equity; and public sector management and institutions.

Given the importance of the local context for the outcomes of decentralization and of other models of service delivery, reliable indicators of accountability and governance are essential for effective policy, program, and project design. That is, these indicators provide a picture of a country's accountability and governance, which helps tailor policies and reforms to its context and achieve better results. Reliable performance indicators that measure progress in outputs, outcomes, and environmental impact are also essential for a robust monitoring framework, especially in the scope of OBA or other RBF approaches.

[20] UNDP (2007) is a very useful user guide that covers 35 governance indicator sources, describing the main characteristics of each source and providing guidance on how to use it.

[21] Reports aggregate and individual governance indicators for 215 economies, over the period 1996–2012, for six areas of governance. Data available at: http://www.govindicators.org.

[22] Overall, the ratings focus on the quality of each country's current policies and institutions, which are the main determinant of present aid effectiveness prospects. The importance of the CPIA in the IDA Performance- Based Allocations is underscored by the fact that the overall country score is referred to as the IDA Resource Allocation Index (IRAI). IDA's resources are allocated on the basis of Country Performance Ratings (CPR), which consist of a linear combination of the 4 clusters of the CPIA ratings (joint weight of 92 %) with the portfolio performance rating (weight of 8 %). Data available, for the period 2005–2012, at: http://databank.worldbank.org/data/databases.aspx.

5 Lessons from the Recent International Experience and Implications for the Nexus Approach to Management of Environmental Resources

Although many lessons can be learned from experience, what works (and what does not work) is by no means a settled matter. Decentralization and other alternatives to centralized public delivery create both opportunities and potential problems. To take advantage of the former while avoiding the latter, several institutional and incentive elements should be present. The main lessons from the international experience and implications for the nexus approach to the management of environmental resources are:

- The increase in local government political, administrative, and financial discretion must be accompanied by adequate degrees of accountability in those three dimensions. This requires policies, laws, environmental standards, and institutions that create incentives for good governance and reduce the need for costly monitoring and enforcement action by the central government. The social/political/economic context should be carefully studied and taken into consideration in service delivery reforms, as solutions that work well in some countries may prove to be disastrous in others.
- Active local participation and engagement of citizens is another very important determinant of the success or failure of decentralization and of alternative service delivery models, and of the proper implementation of the nexus approach to the management of environmental resources, as it is essential to ascertain consumers' views on the reliability, affordability, and adequacy of services.
- While there is evidence that private participation in service delivery leads to greater operational efficiency and reliability of services, concern for the poor and for the environment is by no means automatic. Thus, leases, concessions, BOTs, BOOs, and other PPP arrangements need to include enforceable requirements that affordable services will be available to all sections of society. It is also necessary to assure that environmental sustainability is maintained.
- Given the relatively small volume of private investment in the water and sanitation sector, PPPs should be seen as a way of improving operational efficiency and service quality rather than as a major source of finance. Additionally, solid legal and policy frameworks are essential to the attraction of private investment, as they play an important role in guaranteeing that costs will be recovered.
- OBA and RBF in general have the potential to allocate donor funds more efficiently and to increase transparency, accountability, and efficiency in service delivery. But, existing approaches need to be applied with greater care and new results-based approaches should not be applied unless they respond to the demands of recipient country governments in consultation with civil society and parliaments, and have been designed to meet aid effectiveness and resource sustainability principles. Well-defined required outputs/results, which lead to a

proper management of water, waste, and soil, can certainly help advance the nexus approach.

- Especially in countries that have made substantial progress in sector reform, it makes sense to move forward from sector-wide approaches (SWAPs) to budget support, as the latter allows for a broader development perspective prioritizing investment requirements and sector and regional priorities. This broader approach, based on a medium-term programmatic setting, provides a way of dealing with questions of interactions between legal and policy reform, structural changes in the economy and society and variability in the biophysical environment. Thus, it is an appropriate framework for the advancement of the nexus approach.
- Capacity building, especially to increase skills and accountability, and facilitate technology adaptation, among local government units, NGOs, and communities is a necessary ingredient of the sustainable and integrated management of water, soil, and waste resources. According to Kurian and Ardakanian (2014), the successful design, implementation, and evaluation of results-based financing strategies require the development of capacities for transdisciplinary approaches to planning and management of environmental resources. In this sense, institutions such as UNU-FLORES play a very important role in providing blended learning curricula for the training of policymakers, administrations, and members of NGOs and local communities in the integrated management of water, soil, and waste.
- Reinforcing accountability of policymakers and building capacity of politicians and stakeholders are essential to address questions satisfactorily of intersectionality among material fluxes, public financing, heterogeneity, and changes in institutional and biophysical environment, relevant to the nexus approach to environmental management.
- Indicators/indices play an important role in policy, program, and project design and in monitoring results. Although there are already several databases of accountability/governance indicators, they do not provide subnational data. Reliable indicators are also essential to program evaluation and monitoring, especially in the scope of OBA and other types of RBF. There is considerable room for improvement in this area, and it is vital to use indicators that actually record intended results and do not create perverse incentives. Finally, the implementation of the nexus approach to the management of environmental resources and the elaboration of quantitative trade-off analyses require the development of indicators that address sustainable resource use, human well-being and equity, as well as integrated assessments of water, energy and food sectors (Kurian and Ardakanian 2014). Scientifically robust analyses based on those indicators are necessary to avoid tensions, monitor results and inform policies and regulations across the nexus.

Keywords and Definitions

Accountability	Obligation to explain and justify conduct. Implies a relationship between a principal (or account holder) and an agent (or accountee) in which the former is able to keep in check the actions of the latter
Budget support	Budget support operations channel funds to national budgets and engage with fundamental policy reforms
Capacity building	Enhancement and development of public officials and citizens' capabilities
Discretion	The capacity to decide on political, administrative, and financial matters. Public officials have discretion to use their best judgement in applying rules and policies to meet the public interest
Governance	The manner in which the state acquires and exercises its authority to provide public goods and services
Public-private partnership	Government service or private business venture, which is funded and operated through a partnership of government and one or more private sector companies
Results-based financing	Funding model in which payments are only made after previously agreed upon results/outcomes can be proven to have been obtained

References

Agrawal A, Ribot J (1999) Accountability in decentralization: a framework with South Asian and African cases. Journal of Developing Areas 33:473–502

Allen A, Dávila JD, Hofmann P (2006) Governance of water and sanitation services for the peri-urban poor: a framework for understanding and action in metropolitan regions. Development Planning Unit—UCL, London

Annez PC (2006) Urban infrastructure finance from private operators: what have we learned from recent experience? World Bank Policy Research Working Paper 4045

Ardanáz M, Leiras M, Tommasi M (2014) The politics of federalism in Argentina and its implications for governance and accountability. World Dev 53:26–45

Arze del Granado F, Martinez-Vazquez J, McNab R (2005) Fiscal decentralization and the functional composition of public expenditures. Working paper 05-01, Andrew Young School, Georgia State University

Baltaci M, Yilmaz S (2006) Keeping an eye on subnational governments. Internal control and audit at local levels. World Bank Institute Working Paper Series No. 37257. The World Bank, Washington, DC

Bardhan P, Mookherjee D (2000) Capture and governance at local and national levels. Am Econ Rev 90(2):135–139

Blair H (2000) Participation and accountability at the periphery: democratic local governance in six countries. World Dev 28(1):21–39
Boadway R, Shah A (2009) Fiscal federalism: principles and practice of multiorder governance. Cambridge University Press, Cambridge
Brennan G, Buchanan J (1980) The power to tax: analytical foundations of a fiscal constitution. Cambridge University Press, Cambridge
Davis J (2004) Corruption in public service delivery: experience from South Asia's water and sanitation sector. World Dev 32(1):53–71
De Grauwe A, Lugaz C, Baldé D, Diakhaté C, Dougnon D, Moustapha M, Odushina D (2005) Does decentralisation lead to school improvement? Findings and lessons from research in West Africa. J Educ Int Dev 1(1)
Diamond L, Linz JJ, Lipset SM (1995) Politics in developing countries: comparing experiences with democracy. Lynne Reinner, Boulder
Dinar A, Kemper K, Blomquist W, Kurukulasuriya P (2007) Whitewater: decentralisation of river basin water resource management. J Policy Model 29(6):851–867
Díaz-Cayeros A, Magalosi B, Ruiz-Euler A (2014) Traditional governance, citizen engagement, and local public goods: evidence from Mexico. World Dev 53:80–93
Faguet JP (2014) Decentralization and governance. World Dev 53:2–13
Fan CS, Lin C, Treisman D (2009) Political decentralization and corruption: evidence from around the world. J Public Econ 93:14–34
Fisman R, Gatti R (2002) Decentralization and corruption: evidence across countries. J Public Econ 83:325–345
Golooba-Mutebi F (2005) When popular participation won't improve service provision: primary health care in Uganda. Dev Policy Rev 23(2):165–182
Gómez EJ (2008) A temporal analytical approach to decentralisation: lessons from Brazil's health sector. J Health Polit Policy Law 33(1):53–91
Gonçalves S (2014) The effects of participatory budgeting on municipal expenditures and infant mortality in Brazil. World Dev 53:94–110
Hechter M (2000) Containing nationalism. Oxford University Press, New York
IDA (2009) A review of the use of output-based aid approaches. International Development Association, Washington, DC
Ivanyna M, Shah A (2011) Decentralization and corruption: new cross-country evidence. Environ Plann C: Gov Policy 29:344–362
Ivanyna M, Shah A (2013) How close is your government to its people? Worldwide indicators on localization and decentralization. Economics Discussion Paper No. 2013-38
Iyer P, Evans B, Cardosi J, Hicks N (2005) Rural water supply, sanitation and budget support: guidelines for task teams. World Bank, Washington, DC
Jones N, Lyytikainen M, Mukherjee M, Reddy MG (2007) Decentralisation and participatory service delivery. J Child Poverty 13(2):207–229
Jütting J, Kauffmann C, McDonnell I, Osterrieder H, Pinaud N, Wegner L (2004) Decentralisation and poverty in developing countries: exploring the impact. OECD Development Centre Working Paper 236. OECD, Paris
Kurian M (2010) Financing the millennium development goals (MDGs) for water and sanitation: issues and options. In: Kurian M, McCarney P (eds) Peri-urban water and sanitation services: policy, planning and method. Springer, Dordreth, pp 133–154
Kurian M, Ardakanian R (2013) Institutional arrangements and governance structures that advance the nexus approach to management of environmental resources. White book—Advancing a nexus approach to the sustainable management of water, soil and waste. UNU-FLORES. Chapter 4, pp. 56–66
Kurian M, Ardakanian R (2014) Introduction: the nexus approach to environmental governance considering global change. UNU—Springer, forthcoming
Kurian M, Dietz T (2013) Leadership on the commons: wealth distribution, co-provision and service delivery. J Dev Stud 49(11):1532–1547

LDI (2013) The role of decentralisation/devolution in improving development outcomes at the local level: review of the literature and selected cases. Local Development International LLC, New York

Litvack J, Seddon J (eds) (1999) Decentralization briefing notes. WBI Working Papers. World Bank Institute, Washington, DC

Marin P (2009) Public-private partnerships for urban water utilities: a review of experiences in developing countries. The World Bank, Washington, DC

Martinez-Vazquez J, Yao MH (2009) Fiscal decentralization and public sector employment: a cross-country analysis. International Center for Public Policy Working Paper Series, at AYSPS, GSU paper 0903

Myerson R (2014) Constitutional structures for a strong democracy: considerations on the government of Pakistan. World Dev 53:46–54

Nelson MA (1987) Searching for Leviathan: comment and extension. Am Econ Rev 77 (1):198–204

Oates W (1985) Searching for Leviathan: an empirical study. Am Econ Rev 75(4):748–757

OECD (2009) Strategic financial planning for water supply and sanitation. Environment Department, Paris

Ostrom E (1990) Governing the commons: the evolution of institutions for collective action. Cambridge University Press, Cambridge

Pereira J, Villota C (2012) Hitting the target? Evaluating the effectiveness of results-based approaches to aid. European network on debt and development—Eurodad, Brussels

Shleifer A, Vishny RW (1993) Corruption. Q J Econ 108(3):599–617

Treisman D (2000) The causes of corruption: a cross-national study. J Public Econ 76(3):399–457

UNDP (2007) Governance indicators: a user's guide, 2nd edn. United Nations Development Programme, New York

Weingast BR (2014) Second generation fiscal federalism: political aspects of decentralization and economic development. World Dev 53:14–25

WELL (1998) Guidance manual on water supply and sanitation programmes. Water, Engineering and Development Centre (WEDC)-DFID, London

World Bank (2004) World development report: making services work for the poor. The World Bank, Washington, DC

World Bank (2007a). Local government discretion and accountability: a local governance framework. Report No. 40153. The World Bank, Washington, DC

World Bank (2007b) Strengthening World Bank group engagement on governance and anticorruption. The World Bank, Washington, DC

Yilmaz S, Beris Y, Serrano-Berthet R (2010) Linking local government discretion and accountability in decentralisation. Dev Policy Rev 28(3):259–293

Zax JS (1989) Is there a Leviathan in your neighborhood? Am Econ Rev 79(3):560–567

MIX
Papier aus verantwortungsvollen Quellen
Paper from responsible sources
FSC® C105338
FSC
www.fsc.org

If you have any concerns about our products,
you can contact us on
ProductSafety@springernature.com

In case Publisher is established outside the EU,
the EU authorized representative is:
Springer Nature Customer Service Center GmbH
Europaplatz 3, 69115 Heidelberg, Germany

Printed by Libri Plureos GmbH
in Hamburg, Germany